Global Advances in Engineering Education

Industrial and Systems Engineering Series

Series Editors:

Waldemar Karwowski

University of Central Florida, Orlando, USA

Hamid R. Parsaei

Texas A&M University, College Station, USA

Industrial Engineering has evolved as a major engineering and management discipline, the effective utilization of which has contributed to our increased standard of living through increased productivity, quality of work, and quality of services and improvements in the working environments. The Industrial and Systems Engineering book series provides timely and useful methodologies for achieving increased productivity and quality, competitiveness, globalization of business and for increasing the quality of working life in manufacturing and service industries. This book series should be of value to all industrial engineers and managers, whether they are in profit motivated operations or in other nonprofit fields of activity.

Manufacturing Productivity in China

Edited by Li Zheng, Simin Huang, and Zhihai Zhang

Human Factors in Transportation: Social and Technological Evolution Across Maritime, Road, Rail, and Aviation Domains

Giuseppe Di Bucchianico, Andrea Vallicelli, Neville A. Stanton, and Steven J. Landry

Laser and Photonic Systems: Design and Integration

Edited by Shimon Y. Nof, Andrew M. Weiner, and Gary J. Cheng

Supply Chain Management and Logistics: Innovative Strategies and Practical Solutions

Edited by Zhe Liang, Wanpracha Art Chaovalitwongse, and Leyuan Shi

For more information about this series, please visit: https://www.crcpress.com/Industrial-and-Systems-Engineering-Series/book-series/CRCINDSYSENG

Global Advances in Engineering Education

Edited by
J. P. Mohsen, Mohamed Y. Ismail,
Hamid R. Parsaei, and Waldemar Karwowski

CRC Press is an imprint of the
Taylor & Francis Group, an **informa** business

CRC Press
Taylor & Francis Group
6000 Broken Sound Parkway NW, Suite 300
Boca Raton, FL 33487-2742

First issued in paperback 2021

ISBN-13: 978-0-367-77978-8 (pbk)
ISBN-13: 978-1-138-05190-4 (hbk)

Library of Congress Cataloging-in-Publication Data

Names: Mohsen, J. P., editor. | Ismail, Muhammad Y., 1985- editor. | Parsaei, Hamid R., editor. | Karwowski, Waldemar, 1953- editor.
Title: Global advances in engineering education / edited by J.P. Mohsen, Mohamed Y. Ismail, Hamid R. Parsaei, and Waldemar Karwowski.
Description: Boca Raton, FL : CRC Press/Taylor & Francis Group, 2019. | Series: Industrial and systems engineering series | Includes bibliographical references.
Identifiers: LCCN 2018059544 | ISBN 9781138051904 (hardback : acid-free paper) | ISBN 9781315168074 (ebook)
Subjects: LCSH: Engineering—Study and teaching.
Classification: LCC T65 .G56 2019 | DDC 620.0071—dc23
LC record available at https://lccn.loc.gov/2018059544

Visit the Taylor & Francis Web site at
http://www.taylorandfrancis.com

and the CRC Press Web site at
http://www.crcpress.com

Contents

Preface

Engineering education either in its classical training forms or more traditional hands-on and apprenticeship methods has been around for over several thousand years. Engineers by and large are credited for creating public infrastructures which in turn lead into national prosperity and advancement of communities. Academic education to train competent engineers has been the subject of numerous books, articles, and global technical conventions and symposia. The need to produce well-trained and properly educated engineers with vast problem-solving and communication skills with clear appreciation of their societal and ethical responsibilities has been the subject of many academic debates. Engineering trainings regardless of their focus and specializations have always been publicly perceived as demanding and often require strong background in mathematics and basic sciences.

Over the past five decades with the advent of space discovery and the development and introduction of mobile and wireless communications, engineering fields in general have achieved significant public recognition. To encourage more high-school students to attend engineering and science-driven fields, several STEM (Science, Technology, Engineering, and Mathematics) initiatives have globally been introduced. To make STEM more attractive and accessible to new generations, new delivery methods such as "flipped classrooms," online and web-based instructions were introduced and their effectiveness in retaining the information has been investigated by many researchers. Historically, research on engineering education has been carried out by social scientists; however, over the past three decades the engineering education research has rightfully earned its position among applied science disciplines and many research journals reported such research progress.

This edited volume is intended to report on some of initiatives and activities undertaken and the outcomes achieved by some academic researchers in several different counties from around the globe. Although the chapters on various issues on engineering education, they collectively report on the advancements made and results achieved by different approaches adopted by several academic practitioners to create a new paradigm for educating and training new engineers. Although developing and implementing creative ways to attract and train new generation of engineers is a dynamic process, sometime a small subset gets reported in books and literature. We are confident the journey continues and more creative approaches will get documented in future publications.

J. P. Mohsen, Ph.D.
Mohamed Y. Ismail, Ph.D.
Hamid R. Parsaei, Ph.D., P.E.
Waldemar Karwowski, Ph.D., P.E.

Editors

J. P. Mohsen, Ph.D., is Past President of the American Society for Engineering Education. He served as ASEE President during 2009–2010. He is the current Vice Chair of the Committee on Education in Engineering of the World Federation of Engineering Organizations (WFEO).

J. P. Mohsen is Professor and Associate Dean of Administration and Faculty Affairs at J. B. Speed School of Engineering, University of Louisville. Prior to that, he served as Chair of the Civil and Environmental Engineering Department. He has been actively involved in the American Society of Civil Engineers, serving on the Educational Activities and Continuing Education committees as well as the Technical Council for Computing and Information Technology. He is currently a member of the Formal Engineering Education Committee of ASCE Council on Sustainability. He is also a member of the Surveying Education Committee of The Utility Engineering and Surveying Institute.

Dr. Mohsen was named Engineer of the Year in Education by the Kentucky Section of ASCE in 1999 and received the University of Louisville Distinguished Service to the Profession Award in 1999 and the Distinguished Teaching Professor Award in 2003. In 2014, he received the ASCE Computing in Civil Engineering Award. He is a Fellow of ASEE as well as a Fellow of ASCE.

Mohamed Y. Ismail, Ph.D., received his MS and Ph.D. in Electrical and Computer Engineering from University of Florida. He progressed through several leadership positions at Verizon Communications. Dr. Ismail joined Texas A&M University at Qatar as a Senior IT Consultant and created several pioneering initiatives for promoting the use of digital technologies in the classroom. He has widely presented papers in national and international conferences and symposia and worked closely with engineering faculty at Texas A&M University at Qatar to develop and introduce new technologies to assist students' learning. His research interests include application of Virtual and Augmented Reality to engineering instruction and the design of innovative instructional methods for engineering education.

Hamid R. Parsaei, Ph.D., P.E., is an internationally recognized leader in the field of engineering education, manufacturing systems design, economic decision making with applications to advanced manufacturing systems, and leadership with more than 34 years of experience. He served as Associate Dean for Academic Affairs, Director of Academic Outreach, and Professor of Mechanical Engineering at Texas A&M University at Qatar while holds the rank of Professor in the Industrial and Systems Engineering and Mechanical Engineering at Texas A&M University in College Station. Dr. Parsaei also served as Professor and Chair of the Industrial Engineering at University of Houston for ten years.

Dr. Parsaei has been principal and co-principal investigator on projects funded by NSF, Qatar Foundation, the US Department of Homeland Security, NIST, NIOSH, Texas DoT, among others, with total funding in excess of $26 million. He has authored or co-authored more than 280 refereed publications in archival journals and conference proceedings. He has held several key leadership positions with the Institute of Industrial and Systems Engineers (IISE). Dr. Parsaei is also a Fellow of the Institute of Industrial and Systems Engineers (IISE), American Society for Engineering Education (ASEE), Industrial Engineering and Operations Management Society International (IEOM), and Society of Manufacturing Engineers (SME).

Dr. Parsaei is a registered professional engineer in the state of Texas.

Waldemar Karwowski, Ph.D., D.Sc., P.E., is Pegasus Professor and Chairman, Department of Industrial Engineering and Management Systems and Executive Director, Institute for Advanced Systems Engineering, University of Central Florida, Orlando, USA. He holds an M.S. (1978) in Production Engineering and Management from the Technical University of Wroclaw, Poland, and a Ph.D. (1982) in Industrial Engineering from Texas Tech University. He was awarded D.Sc. (dr habil.) in management science by the Institute for Organization and Management in Industry, Poland (2004). Dr. Karwowski served on the Committee on Human Factors/Human Systems Integration, National Research Council, the National Academies, USA (2007–2011). Dr. Karwowski currently serves as Co-Editor-in-Chief of Theoretical Issues in Ergonomics Science journal (Taylor & Francis, Ltd), and Editor-in-Chief of Human-Intelligent Systems Integration journal (Springer). Dr. Karwowski has over 500 publications focused on mathematical modeling, soft computing, and computer simulation with applications to the design, analysis and evaluation of complex human–systems integration efforts, including systems engineering, human-centered-design, human performance, safety, neuro-fuzzy systems, nonlinear dynamics and chaos, and neuroergonomics.

Contributors

Abdullah Bafail
King Abdulaziz University
Jeddah, Saudi Arabia

Kailash M. Bafna
Western Michigan University
Kalamazoo, Michigan

Xuemin Chen
Texas Southern University
Houston, Texas

Khaled S. El-Kilany
Arab Academy for Science, Technology, and Maritime Transport
Alexandria, Egypt

Aziz Ezzat El-Sayed
Arab Academy for Science, Technology, and Maritime Transport
Alexandria, Egypt

Mona Enell-Nilsson
University of Vaasa
Vaasa, Finland

Mohamed Y. Ismail
Texas A&M University at Qatar
Doha, Qatar

Minna-Maarit Jaskari
University of Vaasa
Vaasa, Finland

Konstantinos Kakosimos
Texas A&M University at Qatar
Doha, Qatar

Jussi Kantola
University of Vaasa
Vaasa, Finland

Qianlong Lan
University of Houston
Houston, Texas

Maximilian Moll
Universität der Bundeswehr München
München, Germany

Atsuo Murata
Okayama University
Okayama, Japan

Marian Sorin Nistor
Universität der Bundeswehr München
München, Germany

Hamid R. Parsaei
Texas A&M University at Qatar, Education City
Doha, Qatar

Stefan W. Pickl
Universität der Bundeswehr München
München, Germany

Radhey Sharma
West Virginia University
Morgantown, West Virginia

Gangbing Song
University of Houston
Houston, Texas

Osman Taylan
King Abdulaziz University
Jeddah, Saudi Arabia

Ning Wang
University of Houston
Houston, Texas

chapter one

History of engineering education

Aziz Ezzat El-Sayed and Khaled S. El-Kilany
Arab Academy for Science, Technology, and Maritime Transport

Contents

1.1 Prologue: Engineering and the cradle of civilization

With the existence of the first man on earth, his survival was very much linked, dependent on, and correlated to his efforts and success on how to adapt and conquer his surrounding environment and the unfriendly natural phenomena. Thus, he always strived to apply the knowledge he gained to use his physical power, the materials, and other resources around him to make his life easier. Engineering is defined, according to the Encyclopedia Britannica, as "the application of science to the optimum conversion of the resources of nature to the uses of humankind." If we agree on such a generic definition, we may discover that engineering as a science and art, and as we know it today, actually was, implicitly and partially, there from the beginning of written history.

To elaborate more on the meaning and spread of the term "engineering," we may extend it to encompass the application of mathematics, science, economics, experimentation, social, and practical knowledge to design, construct, maintain, and improve structures, machines, systems, components, tools, materials, processes, and organizations.

It is hard to state, however, where and when engineering started historically. Was it near the banks of the great Nile River where the ancient Egyptians built their gigantic temples, like the one shown in Figure 1.1, and the great cities like Thebes (Luxor) and Memphis (near south of modern Cairo) some 3,000 years BC? Was it in the valley of

Figure 1.1 Entrance to the Luxor Temple in Egypt. (By MusikAnimal—Own work, CC BY-SA 4.0, https://commons.wikimedia.org/w/index.php?curid=41830942.)

Tigris and Euphrates rivers, this vast area which was historically known as Mesopotamia (most of current Iraq, plus smaller parts of Syria, Iran, Turkey, and Kuwait), where Sumerians and Akkadians (ancestors of Iraqis) gave birth to the world's first cities, and the cuneiform scripts the predecessor of most of the ancient world alphabets?

Alternatively, was it in the same region where the invention of the wheel as a circular element intended to rotate on an axle bearing, led to the world's first transportation device? Or was it where ancient Babylonians have once established the legendary hanging gardens of Babel shown in Figure 1.2, to be one of the Seven Wonders of the old world? Or was it where the Phoenicians gave birth to other alphabets in Lebanon, during almost the same time, and drove their ships in the Mediterranean Sea as clever and experienced sailors?

Ever since those ancient epochs, if we want to follow the steps of engineering education and how it evolved, we have to assume that as long as there was a lasting monument, a durable structure, and an enduring construction, there should be a form of engineering science. That form of science has been transferred systematically to and practiced by the workforce, foremen, supervisors, designers, and executives who erected those shrines along history. One may conclude that to have, for instance, an everlasting construction as those existed over centuries, this required precise measurement, materials science, metallurgy for hard tools, organization

Figure 1.2 Hand-colored engraving of the legendary Hanging Gardens of Babylon, with the Tower of Babel in the background. (By Maarten van Heemskerck—http://www.plinia.net/wonders/gardens/hgpix1.html, Public domain, https://commons.wikimedia.org/w/index.php?curid=65909.)

structure, material handling equipment, and sometimes knowledge of trigonometry and principles of mechanics.

In this chapter, the analysis of the history of engineering education is actually a study of the evolution of physics, chemistry, philosophy, geometry, astronomy, mathematics, and logic sciences, because all of these disciplines were practiced by most of the men of wisdom in ancient times. Moreover, these forms of human knowledge present the backbone of engineering as we know it today. This chapter also focuses on the main contributions in fields related to engineering and not the biographies neither of the developers nor the political leaders. It presents a chronological study of the achievements of the world from the perspective above. The study marked the pioneering efforts, brilliant contributions, and breakthrough inventions. It is not intended to be encyclopedic in nature, which means that some of the places, dynasties, and engineering achievements would not be covered due to the limited space allocated to this chapter. Nevertheless, we believe that such an issue could be widely discussed and it could be the subject matter of a more comprehensive work. It is rather impossible to elaborate on every engineering and engineering education aspects in history. However, we have tried to mention, from the previously mentioned perspective, the first recorded effort and contribution across the history of humanity. Since our focus is history, we care much to mention the questions of when, where, and in what discipline was the original contribution. We were not concerned about what is the current status of that old or ancient innovation at least that was the priority we try to adhere to. Did those criteria make sense? It is the reader who will figure out the answer to the aforementioned question.

1.2 *Engineering in ancient Egypt*

We start with engineering in ancient Egypt thousands of years BC, where no schools or universities exist at that time. However, one could not imagine that a megastructure and a super project like the great pyramid of Giza, which is branded as the pyramid of Khufu (also known as Cheops), was designed and built without an engineering science and background. Exactly how it was built, what was the true reason for constructing such an enormous building, still puzzles people today. Although known as a multitalented and iconic historical figure, Imhotep (about 2,600 BC), who devised a means of creating the step pyramid of Saqqara for his king Dzjoser (Figure 1.3), has never been mentioned to have an engineering background. Moreover, there is no record that the chief designer and architect of the great pyramid did receive any formal engineering education like what we know nowadays.

Figure 1.3 The Step Pyramid of Dzjoser at Saqqara, Egypt. (By Dennis Jarvis from Halifax, Canada—Egypt-12B-021-Step Pyramid of Djoser, CC BY-SA 2.0, https://commons.wikimedia.org/w/index.php?curid=66938816.)

The secret of mastering those ancient stunning engineering achievements is still a mystery, but we may observe that the knowledge behind such scientific miracles was not recorded on the walls of the temples or found in the buried tombs discovered beneath the hot sands of the deserts.

One may assume that there was science, theories, mathematics, and geometrical advances which form the bases and foundation for such incredible monuments.

Consequently, the direct conclusion is that there was an engineering education, in a way or form that explains the aforementioned huge outcomes. We may also conclude that inside the sacred temples of ancient Egypt, and similar holy places, such teachings were in the hands of the priests of Amen-Ra who were the sole source of knowledge, wisdom, and sometimes political power.

The only explanation for not being revealed is that it was an art, science, and knowledge that should not be disseminated, publicized, or transferred by any means because it was a source of power and it was closely related to their religion and divine beliefs.

Ancient Egyptians strongly believed in the idea of immortality derived from religious roots. This idea was the main reason behind the selection of very hard building materials (e.g., granite) to build their temples and tombs and to discover appropriate technologies to cut, shape, transport, and install massive amounts of stones.

1.3 Academy of Athens, Greece

During the 8th-century BC, the Greeks started to use the Phoenician (ancestors of the Lebanese) alphabet and adapted it to their own language, creating the first "true" alphabet. A process was needed to establish the Greek's unique contribution to human scientific knowledge heritage. As one of the oldest in the world, Plato established his Academy in Athens, Greece around 387 BC (Figure 1.4). Aristotle (384–322 BC) studied there for 20 years before founding his own school, the Lyceum. The Academy persisted throughout the Hellenistic period as a skeptical logic school. Among the attendees of such an Academy "Akademia" was Theaetetus of Sunium (417–368 BC), a mathematician whose principal contributions were on irrational lengths, which was included in Euclid's *Elements*. His teacher, Theodorus of Cyrene, had explored the theory of incommensurable quantities. Archytas (428–347 BC) was a philosopher, mathematician, and astronomer. He was also a scientist of the Pythagorean school (due to Pythagoras 570–495 BC) and famous for being the reputed founder of mathematical mechanics. Archimedes of Syracuse (287–212 BC), as shown in Figure 1.5, was a mathematician, physicist, engineer, inventor, and astronomer. He is considered one of the leading scientists in the ancient world and one of the greatest of all time. Archimedes anticipated

Figure 1.4 "Scuola di Atene," The School of Athens. (By Raphael—Stitched together from vatican.va, Public Domain, https://commons.wikimedia.org/w/index.php?curid=4406048.)

Figure 1.5 Archimedes Thoughtful by Fetti (1620). (By Domenico Fetti—http://archimedes2.mpiwg-berlin.mpg.de/archimedes_templates/popup.htm, Public Domain, https://commons.wikimedia.org/w/index.php?curid=146592.)

modern calculus and analysis by applying concepts of infinitesimals and the method of exhaustion to derive and rigorously prove a range of geometrical theorems, including the area of a circle, the surface area and volume of a sphere, and the area under a parabola. The Archimedes' screw is still in use today for pumping liquids and granulated solids such as coal and grain. Autolycus (360–290 BC) was an astronomer, mathematician, and geographer. The lunar crater Autolycus was named in his honor.

The Academy of Athens had its strong influence on the lifestyle of ancient Greek people and it shaped the free-thinking environment at that time. The result was a great evolution in mathematics, philosophy, geometry, astronomy, and other facets of knowledge which, as already stated, created the backbone of any formal engineering education and consequently accomplishments.

As a model, the Parthenon shown in Figure 1.6 is an ancient temple on the Acropolis, Greece, which was constructed between 447 and 432 BC. It represents the most important remaining and enduring building of ancient classical Greek architecture and art. Callicrates and Ictinus, who

Figure 1.6 The Parthenon showing the common structural features of ancient Greek architecture. (By Steve Swayne—File:The Parthenon Athens.jpg Wikimedia Commons, CC BY-SA 2.0, https://commons.wikimedia.org/w/index.php?curid=22556135. From Travelers in the Middle East Archive (TIMEA). http://hdl.handle.net/1911/9303. By Antonio Tempesta (Italy, Florence, 1555–1630), Public Domain, https://commons.wikimedia.org/w/index.php?curid=27300728.)

both lived in the middle of the 5th-century BC, were considered to be the first architects of the Parthenon.

1.4 *The city of Alexandria*

Alexander the Great founded the city of Alexandria in 323 BC near the western branch of the Nile on a site between the sea and Lake Marriott, linking the island of Pharos to the mainland of Egypt. The urban plan of the city was set by Alexander's Architect Dinocratis, who designed it to have unique features. Unlike other older cities, Alexandria was remarked by its two straight long streets: one parallel to the Mediterranean Sea shores from east to west and the other a perpendicular crossroad as illustrated in Figure 1.7.

Alexander built other cities carrying his name. Around 15 other "Alexandria" cities were originally founded by Alexander in his vast campaign during the 3rd-century BC. More than double that number of cities carry the same name, "Alexandria," in all continents. However, after more than 23 centuries, Alexandria of Egypt remains the largest and the most famous. On the other hand, and despite its current overwhelming population together with the expansion of the city to the east and west, Alexandria, as its old city version, still carries the same urban design characteristics intended by its founders.

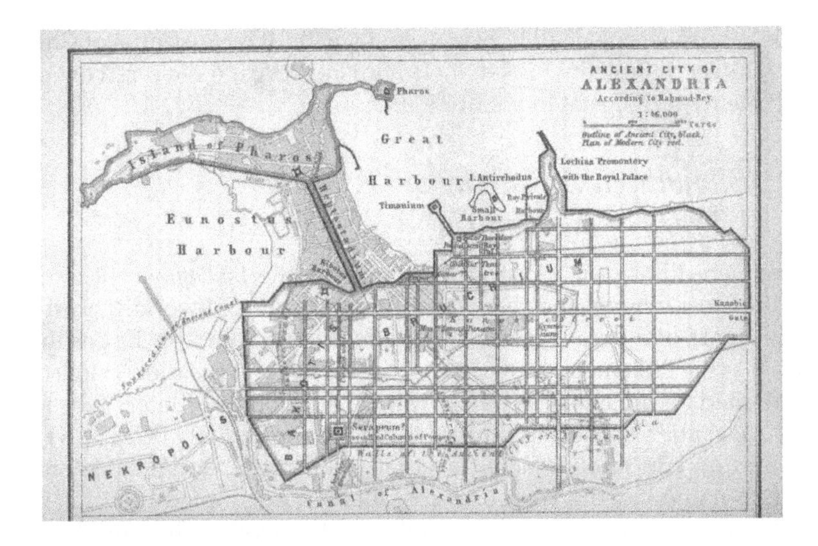

Figure 1.7 Plan of the ancient city of Alexandria.

Figure 1.8 An imaginary illustration of the ancient lighthouse of Alexandria.

Another great engineering achievement was the establishment of the ancient lighthouse of Alexandria, which was located in the western parts of ancient Alexandria and shown in Figure 1.8. It is known as Pharos of Alexandria and is one of the Seven Wonders of the World and the most famous lighthouse in antiquity. It was a technological triumph in its time and is the archetype of all lighthouses since. Built by Sostratus of Cnidus, the architect of Ptolemy I Soter, it was finished during the reign of Soter's son Ptolemy II of Egypt in about 280 BC.

1.5 Bibliotheca Alexandrina "Ancient Library of Alexandria"

The ancient library of Alexandria, known as the *Bibliotheca Alexandrina,* was the most famous and the largest library in the old world. The library formed part of a great research institute at Alexandria in Egypt that is known as the Ptolemaic "Mouseion" Academy. The expression "Mouseion" was slanted to "museum" as we know it today. Ptolomy I "Soter," who ruled after Alexander, started to build the library around 295 BC. He wanted to found an educational facility in the Greek style, similar to Aristotle's institute named as the "Lyceum" in Athens. He assumed that this place would attract great scholars from all over the world. The initial design shown in Figure 1.9 is attributed to Demetrius Phalereus and is estimated to have stored at its peak 400,000–700,000 parchment and papyri scrolls. Bibliotheca Alexandrina attracted many full-time researchers, scientists,

Figure 1.9 An imaginary illustration of the ancient Library of Alexandria, Egypt. (By O. Von Corven—Tolzmann, Don Heinrich, Alfred Hessel and Reuben Peiss. The Memory of Mankind. New Castle, DE: Oak Knoll Press, 2001, Public Domain, https://commons.wikimedia.org/w/index.php?curid=2307486.)

and students for three centuries to create top-notch education at that time. Supported by state stipends that helped them maintain the scrolls, translate and copy them, and conduct research. As time went on, the city opened another branch of the library at the Temple of Serapis.

It should be mentioned that major contributors to mathematics, science, and engineering lived most of their lives in Alexandria. Big Names like Euclid was born and died in Alexandria (323 and 285 BC), often referred to as the "founder of geometry" or the "father of geometry." Apollonius of Perga was a geometer and astronomer known for his theories on the topic of conic sections, died in Alexandria 190 BC. Hero of Alexandria (10–c. 70 AD) was a mathematician and engineer, who was the representative of the Hellenistic scientific tradition. Hero published a well-recognized description of a steam-powered device sometimes called a "Hero engine." Eratosthenes of Cyrene, who died in Alexandria, 194 BC, was known as a scientific writer, astronomer, and the one who made the first measurement of the size of the Earth. Theon of Alexandria was a mathematician and astronomer who is best remembered for the part he played in the preservation of Euclid's Elements. Theon's daughter Hypatia, born and died in Alexandria (355–415), was known as the earliest female mathematician, astronomer, and philosopher. Historically, the vanishing of the ancient library of Alexandria is a mystery. However, it was mentioned that the library had been burned to the ground several times in the early centuries AD.

1.6 Engineering in ancient Rome

For almost a millennium, ancient Rome, as the capital of an empire, controlled the destiny of all civilization extended from what is known as Europe, north, south, and east Mediterranean nations. On the other hand, Rome's early development has been largely influenced by those civilizations in Greece, Carthage, Egypt, Phoenicia, and Persia. The ruins leftover in several countries which carry the features and characteristics of Roman culture, design, and architecture are living examples of how such mutual influence exists. The city of Rome itself holds a vast number of lasting monuments, tributes, temples, and memorials that indicate a sophisticated engineering science and technology even as we recognize it nowadays.

One example is the Colosseum (c. 70–82 CE), shown in Figure 1.10, which was built during the Flavian dynasty of emperors. The oval stadium measures about one-half of a kilometer around, with external dimensions of 190 by 155 m. The approximately 48-m facade has three superimposed series of 80 arches and an attic story. The façade also included sunshades to protect the 50,000 seats from the sun during the gladiatorial contests, combats with wild animals, and false battles.

For more than 1,000 years, the Colosseum was a famous symbol of Rome, and it had been a sort of entertainment for the people of the city. Once again,

Figure 1.10 A 4 × 4 segment panorama of the Colosseum at dusk. (By Diliff—Own work, CC BY-SA 2.5, https://commons.wikimedia.org/w/index.php?curid=2127844.)

an unbiased analysis would conclude that such an unprecedented structure would not be designed, erected, and maintained without a proper kind of engineering and technology-based experience and knowledge which should have been disseminated by an anonymous mentor to his followers. The list of similar engineering contributions includes tens if not hundreds, for example, the Roman contribution in the design and the construction of bridges is enormous. For instance, the several bridges along the Tiber River built about the time of Hadrian (reigned 117–138 AD) (Figure 1.11).

Figure 1.11 Saint Angelo Bridge over the Tiber River, Rome, Italy. (By Jebulon—Own work, CC0, https://commons.wikimedia.org/w/index.php?curid=30172578.)

1.7 Medieval Bagdad and the House of Wisdom

The Abbasid Empire (762–1258) was the wealthiest, luxurious, and dominant political power in its time reaching from the western Mediterranean to India. Enormous wealth had flowed into the new capital of Baghdad since its foundation in 762. Science, philosophy, and arts were very much encouraged during the reign of Al-Maʾmūn who ruled (813–833).

Al-Maʾmūn encouraged the translation of Greek philosophical and scientific works and founded an academy called the House of Wisdom (Dar al-Ḥikmah) to which the translators, most often Christians, were attached. He also imported manuscripts of particularly important works from Byzantium. Examples are the writings of: Diocles' treatise on mirrors, Theodosius's Spherics, Pappus's work on mechanics, Ptolemy's Planisphaerium, and Hypsicles' treatises on regular polyhedra (of Euclid's Elements) were among those translated. Al-Khwārizmī, (780–850), who worked in (Dar al-Hikma), was a mathematician and astronomer. Al-Khwārizmī introduced Hindu–Arabic numerals and the concepts of algebra that his name was the origin of the terms *algorithm* and *algebra*. Other famous scholars during that period were Thābit ibn Qurrah (836–901), a mathematician, astronomer, physician, and philosopher. Also, Abūʾl-Wafāʾ (940–997) and Omar Khayyam (1048–1131) were capable mathematicians who solved the general problem of extracting roots of any desired degree. Geometry also went through extensive development by Ibrāhīm ibn Sinān (909–946), Abū Sahl al-Kūhī (died c. 995), and Ibn al-Haytham. They solved problems involving the pure geometry of conic sections, including the areas and volumes of plane and solid figures formed from them, and also investigated the optical properties of mirrors made from conic sections.

1.8 Engineering in Andalusia (ancient Spain)

Andalusia in Spain became part of the independent Umayyad caliphate of Córdoba, which was founded by ʿAbd al-Raḥmān III in 929. Despite its political instability, scholars have seen the medieval period as the golden age of Andalusia because of its economic prosperity and its brilliant cultural blossoming. Agriculture, mining, and industry flourished as never before. The cities of Córdoba, Sevilla, and Granada became eminent centers of architecture, science, and learning at a time when the rest of Europe was still emerging from the Dark Ages. The Mosque-Cathedral of Córdoba shown in Figure 1.12 and the fortress-palace of the Alhambra in Granada shown in Figure 1.13 were built during this period, and the great Spanish Muslim philosopher Averroës (Ibn-Rushd) was perhaps its leading intellectual figure.

Figure 1.12 Cathedral–Mosque of Córdoba, Spain. (By Michal Osmenda from Brussels, Belgium—Cathedral–Mosque of Córdoba, CC BY 2.0, https://commons.wikimedia.org/w/index.php?curid=24415630.)

Figure 1.13 Alhambra de Granada, Patio de los Leones (Courtyard of the Lions), Spain. (By Allie Caulfield—originally posted to Flickr as 2002-10-26 11-15 Andalusien, Lissabon 075 Granada, Alhambra, CC BY 2.0, https://commons.wikimedia.org/w/index.php?curid=5780894.)

1.9 *Engineering in Medieval Europe*

One area in which engineering made substantial advances was the construction of cathedrals, castles, and other large structures. Cathedrals were built in the Romanesque style (10th and 11th centuries) and later in the Gothic style (12th–16th centuries).

Another aspect in which engineering made significant progress in medieval Europe was the design and construction of sailing vessels. This led to progress in maritime, and naval construction during that time, first by Spain and Portugal and later by England, set the stage for European exploration and colonization in North and South America, Africa, and Australia.

As for engineering education, one may remark that the first recorded university in Europe was in Bologna, Italy in the 9th century. Although specialized only in Roman law at that time, it is considered the oldest university in continuous operation, as well as one of the leading academic institutions in Italy and Europe.

The historic University of Paris (French: Université de Paris) first appeared in the 12th century, The university is often referred to as the Sorbonne (Collège de Sorbonne) founded in 1257, but the university as such was older and was never completely centered on the Sorbonne.

The University of Paris (along with that of the University of Bologna) became the model for all later medieval universities in Europe.

University of Oxford, England, is also one of the world's old universities that were established in the 12th century as a model of the University of Paris which has gained its reputation from its teachings in theology, law, medicine, and liberal arts. However, both universities were considered centers for mathematics and philosophy. Of particular importance in these universities were the Arabic-based versions of Euclid, of which there were at least four by the 12th century. Of the numerous revisions which were made, that of Johannes Campanus (c. 1250; first printed in 1482) was easily the most popular, serving as a textbook for many generations. Studies of what is now called physics were conducted by Thomas Bradwardine, who was active in Oxford, in the 14th century, as one of the first medieval scholars.

After leaving Paris, Roger Bacon conducted his scientific experiments and lectured at Oxford from 1247 to 1257. Bacon was one of several influential scientists at Oxford during the 13th and 14th centuries. Although departments of engineering science were founded lately with the beginning of the 20th century, it is likely that engineering-related topics were taught and studied at Oxford before that time. Physical sciences were taught and studied at Oxford from at least the 17th century.

The start of Cambridge University was also in the 12th century when scholars from Oxford transferred to Cambridge. Cambridge remained fairly insignificant until about the mid of the 17th century when the professorship of mathematics was handed to Isaac Newton. Newton held the chair for over 30 years and gave the study of mathematics a unique position in the university. He was the eminent figure of the scientific revolution of the 17th century. In optics, his discovery of the composition of white light integrated the phenomena of colors into the science of light and laid the foundation for modern physical optics. In mechanics, his three laws of motion, the basic principles of modern physics, resulted in the formulation of the law of universal gravitation. In mathematics, he was the original discoverer of the infinitesimal calculus.

In medieval and Renaissance Germany, a set of universities were established with the same theological model. One may list Heidelberg University (established in 1384), Leipzig University (1409), and University of Rostock (1419) and a few others. A notable remark here is that some of those universities encompass no explicit engineering departments, even in their current versions. However, engineering closely related fields like mathematics, physics computer science, and economics, are existent.

1.10 Engineering in Asia

1.10.1 The Great Wall of China

As one of the Seven Wonders of the World, the Great Wall of China, shown in Figure 1.14, is an outstanding representative of those famous constructions in China. It was built to be a defensive structure to keep intruders

Figure 1.14 The Great Wall of China. (By Photo by © CEphoto, Uwe Aranas, CC BY-SA 3.0, https://commons.wikimedia.org/w/index.php?curid=32396893.)

from entering the mainland. The construction of the Great Wall began in the 3rd century, and the whole construction project lasted for an entire century, through different dynasties. Finally, this mega structure stretches with a total length of 8851.8 km from west to the east of China, and that traverses nine provinces and municipalities.

1.10.2 Forbidden City, Beijing—China

Known as the Imperial Palace Museum, the Forbidden City shown in Figure 1.15 is a magnificent building complex located in the very heart of Beijing. As the symbol of imperial power, which was built during the Ming Dynasty (1406–1420 AD), it is the largest and well-preserved wooden building complex of the world and is considered a distinguished sample of the traditional Chinese palatial architecture.

1.10.3 Other engineering-related contributions in China

Confirmed by archaeological evidence, the earliest cast iron was developed in China by the early 5th-century BC during the Zhou Dynasty (1122–256 BC), the oldest specimens found in a tomb in Jiangsu province.

The earliest evidence of bronze crossbow bolts dates as early as the mid-5th-century BC in ancient China. Even earlier, paper pulp preparation

Figure 1.15 Forbidden City Beijing Shenwumen Gate, China. (By user:kallgan—Own work, CC BY-SA 3.0, https://commons.wikimedia.org/w/index.php?curid=978574.)

processes and the raw material used in making paper were invented during the Han Dynasty in the 2nd-century BC The birth of paper, as we know it today, took place under the Chinese Han Dynasty in AD 105. Ts'ai Lun, a court official, invented a papermaking process which primarily used rags (textile waste) as the raw material with which to make paper.

1.10.4 Taj Mahal, India

The Taj Mahal, a mausoleum complex in Agra, Northern India and shown in Figure 1.16, has been distinguished as the finest example of Indian, Persian, and Islamic blend of architectural styles. The Taj Mahal is also considered to be one of the most beautiful structural compositions and most iconic monuments in the world today. The plans for the complex have been attributed to various architects of the period, though the chief architect was probably Ustad Aḥmad Lahawrī, an Indian of Persian descent in about 1632. More than 20,000 workers were employed from India, Persia, the Ottoman Empire, and Europe to complete the mausoleum and the adjunct buildings with the decoration work continued until at least 1647.

1.10.5 Nalanda, Bihar, India

Now in ruins, Nalanda used to be a thriving center of learning from the 7th-century BC to1200 CE attracting students and scholars from across

Figure 1.16 Taj Mahal with Taj Mosque. (By Biswarup Ganguly, CC BY 3.0, https://commons.wikimedia.org/w/index.php?curid=35528211.)

Figure 1.17 The ruins of ancient Nalanda in Bihar, India. (By Michael Eisenriegler from Vienna, Austria—Nalanda, CC BY 2.0, https://commons.wikimedia.org/w/index.php?curid=37199366.)

the subcontinent and from as far away as Tibet, China, Korea, and Central Asia. Though the ruins shown in Figure 1.17 occupy an area of just approximately 12 ha, the university once occupied a larger area and consisted of meditation halls, classrooms, temples, and dormitories for over 10,000 students and 2,000 teachers.

1.10.6 Blue Mosque, Turkey

The Blue Mosque, as it is popularly known, or Sultan Ahmet Mosque, shown in Figure 1.18 was constructed between 1609 and 1616 during the rule of Ahmed I. It was designed by the architect, Sedefkar Mohmed Agha, an apprentice of "Sinan Agha the Grand Architect" (1488–1588) who was the chief architect and civil engineer for the sultans during his life period.

1.10.7 Persepolis, Persia (ancient Iran)

Persepolis was the capital of the Achaemenid Empire which is the First Persian Empire (ca. 550–330 BC). It is situated 60 km northeast of the city of Shiraz in Fars Province, Iran. The earliest ruins of Persepolis date back to 515 BC. It represents the Achaemenid style of architecture. UNESCO declared the ruins of Persepolis that shown in Figure 1.19, a World Heritage Site in 1979.

Figure 1.18 The Blue Mosque, Istanbul, Turkey. (By Kamiox—Own work, Public Domain, https://commons.wikimedia.org/w/index.php?curid=1331312.)

Figure 1.19 Ruins of Tachara, Persepolis (ancient Persia). (By Hansueli Krapf—File:2009-11-24 Persepolis 02.jpg, CC BY-SA 3.0, https://commons.wikimedia.org/w/index.php?curid=47322612.)

The city includes a 125,000 m^2 terrace, partly artificially constructed and partly cut out of a mountain, with its east side leaning on a mountain. The other three sides are formed by retaining walls, which vary in height. Rising from 5 to 13 m (16–43 ft) on the west side was a double stair. From there, it gently slopes to the top.

1.11 Engineering during the European Renaissance

Viewed as the passage between the medieval and the modern age, the renaissance is a remarkable milestone in the timeline of Europe's history, which took place between the 14th and 17th centuries. It appeared in Italy as a drastic change in culture, arts, science, and other facets of human lifestyle. From Italy, such a phenomenon was extended to the rest of Europe, until the beginning of the industrial revolution era. During the Renaissance, engineering contributions to mankind were significant. Iconic figures such as Leonardo da Vinci (1452–1519 BC), shown in Figure 1.20, and his substantial achievements in arts, architecture, civil engineering, geology, optics, and hydrodynamics are milestones in world engineering history. In addition to his great paintings such as

Figure 1.20 Leonardo Da Vinci. (By Francesco Melzi, Public Domain, https://commons.wikimedia.org/w/index.php?curid=15498000.)

the *Mona Lisa*, regarded as the most famous portrait of a human figure and the Last Supper the most reproduced religious painting of all time, Da Vinci is also known for his technological inventions in a significant number of different engineering-related areas. He conceptualized a flying machine that is illustrated in Figure 1.21, which may be considered the ancestor of the helicopter, and the device was outlined to be built of wood, reeds, and clothing. Leonardo also envisioned a type of armored fighting vehicle, the origin of the same current military gadget. He also drafted a concentrated solar power which is a system to generate solar power by using mirrors or lenses to concentrate a large area of sunlight,

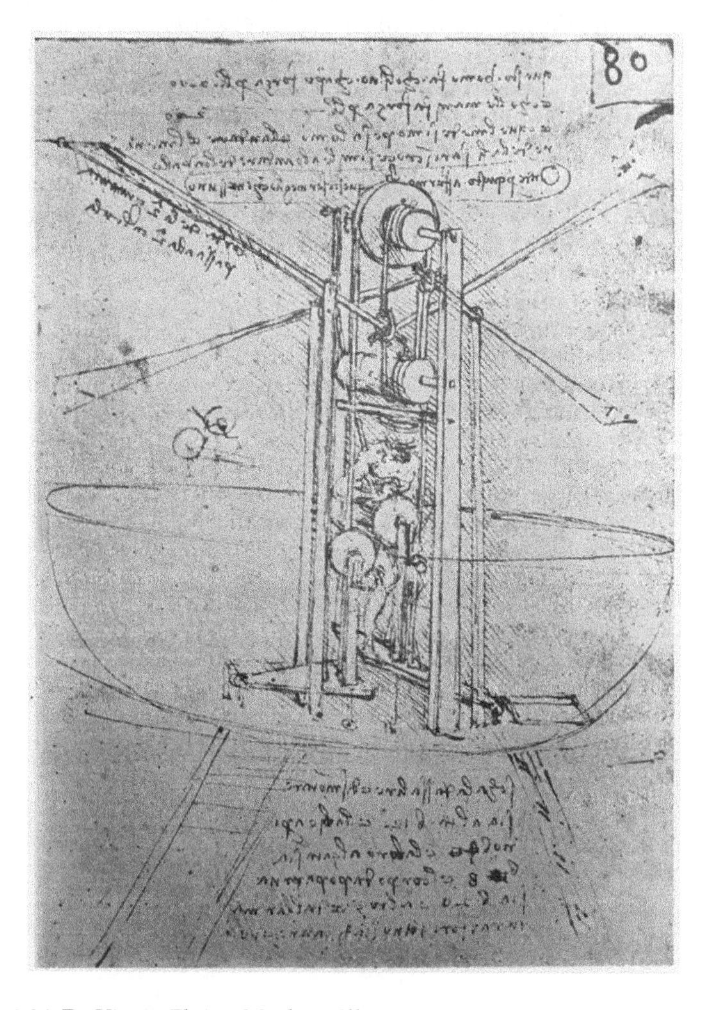

Figure 1.21 Da Vinci's Flying Machine Illustration. (By Leonardo da Vinci, Public Domain, https://commons.wikimedia.org/w/index.php?curid=11228914.)

or solar thermal energy, onto a small area. He drafted thoughts about an adding machine and a double ship hull. Although Da Vinci was an enthusiastic, visionary and productive designer, few of his designs were constructed or even feasible during his lifetime. The reason is that the current advances in manufacturing, metallurgy, construction, and mechanical engineering were not known during his time to materialize his thoughts.

Other significant scientific advances during the renaissance time were made by Galileo Galilei (1564–1642) in astronomy, kinematics, dynamics, physics, engineering, and mathematics.

Also, Tycho Brahe (1546–1601) had significant contribution in astronomy and inventions of accurate instruments of measurement. Meanwhile, Johannes Kepler (1571–1630) discovered the popular laws of orbital and planetary motion. Other examples would include Copernicus (1473–1543), who revolutionary theorized that the Earth moved around the Sun and not the opposite paving the way to real scientific thinking, which consequently led to aeronautical engineering as we know it today.

Another important development was in the process of discovery, focusing on empirical evidence and the importance of mathematics. An early and influential proponent of these ideas was Francis Bacon (1561–1626).

A major contribution to scientific theories, ideas, and logic thinking, which form the backbone of any engineering education, was the invention of printing and the printing press by Johannes Gutenberg (1400–1468). Such a milestone played a key role in the development of the Renaissance, Reformation, the Age of Enlightenment, and the Scientific Revolution.

Consequently, the development of the current engineering disciplines is based principally on that period's influences in physics, chemistry, and mathematics and their extensions into materials science, solid and fluid mechanics, thermodynamics, heat transfer and materials processing, and systems analysis.

1.12 Engineering and the Industrial Revolution

The Industrial Revolution is another landmark in the history of engineering. It is the process causing the dramatic change from a craft-based industry to the machine-based manufacturing. This process began in Britain in the 18th century and later spread to the rest of the world.

Among such technological transformation processes were the substitution of water, wind, human, and animal power by machine power. The invention of a developed version of the steam engine by James Watt (1736–1819), a Scottish mechanical engineer (shown in Figure 1.22), provided the means for progressive development in several different areas. One was the introduction of machines into the manufacturing

Figure 1.22 James Watt (1737–1819). (By John Partridge—https://www.nationalgalleries.org/collection/artists-a-z/b/artist/sir-william-beechey/object/james-watt-1736-1819-engineer-inventor-of-the-steam-engine-pg-2612, Public Domain, https://commons.wikimedia.org/w/index.php?curid=42586342.)

of textiles which was a primitive fabrication process, where people used to make clothes by working in their homes or in small groups. After the introduction of the steam engine, illustrated by James Watt in Figure 1.23, the cloth was made in large factories using machinery powered by steam engines. Thus, textiles engineering with its two major branches, namely the spinning and weaving industry, took place in England and the United States. Further, such complexity introduced the concept of the factory, the system, and the industrial organization to the world. This, in turn, had a dramatic impact on the economy, which led Adam Smith in (1776) to write his famous and fundamental book about "The Wealth of Nations" which restructures the bases for economic concepts and growth.

In addition, the steam engine transformed transportation in particular, the development of the steamship, and the steam locomotive greatly increased the speed with which people could move and increase the number of materials and goods that could be moved. For instance, the formulation above of classical economics by Adam Smith caused the

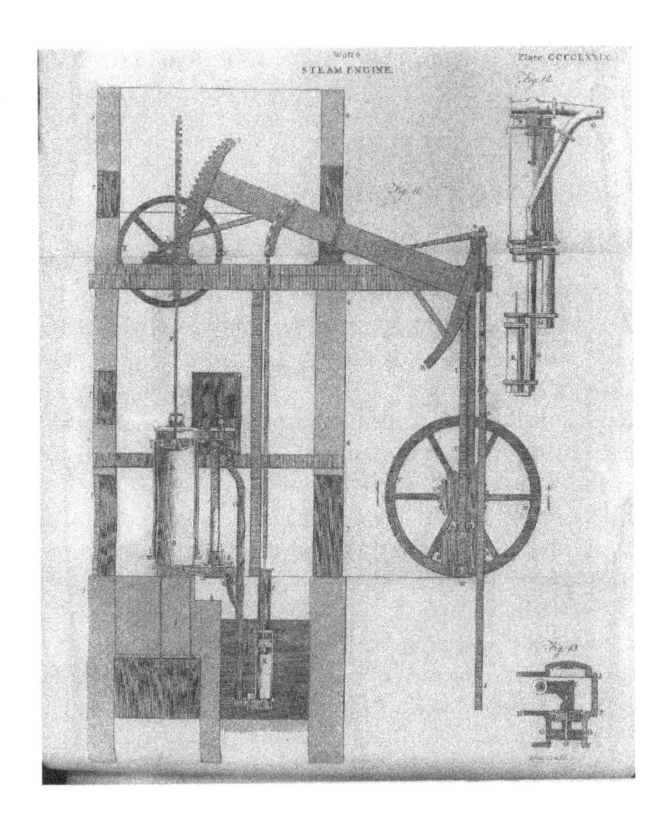

Figure 1.23 A drawing of James Watt's Steam Engine printed in the 3rd edition Britannica 1797. (By DigbyDalton—Encyclopædia Britannica Third Edition (Own work scan), CC BY-SA 3.0, https://commons.wikimedia.org/w/index.php?curid=33043116.)

belief in mercantilism theory of trade to decline. As a result, Britain used its steam-powered naval supremacy to consolidate foreign markets for the resources of the British colonies. The industrial revolution also involved the wide production and use of new basic materials like iron and steel.

Another main feature of the industrial revolution era was the socio-economic changes followed by the invention of telegraph developed between 1830 and 1840 by Samuel Morse (1791–1872) and other inventors; the telegraph revolutionized long-distance communication.

The industrial revolution witnessed the use of other energy sources, including coal, electricity, and petroleum. Along with many contributors during the same period, Samuel Brown, an English engineer and inventor, in 1823 patented the first internal combustion engine to be applied in industry.

These technological changes made possible a tremendously increased use of natural resources and the mass production of manufactured goods.

1.13 Napoleon 1 and the first school of engineering in Europe

One of the first engineering colleges in Europe was the polytechnic college *École Polytechnique* founded in 1794, in France during the times of the French Revolution. It was originally named *École Centrale des Travaux Publics*, Central School of Public Works. Its mission was to provide its students with a scientific background education with a strong emphasis in mathematics, physics, and chemistry, and to prepare them upon graduation to enter the national institutes of public works.

It should be noted that around 50 *École Polytechnique* students, along with a dozen professors and faculty members, accompany General Napoleon Bonaparte (later Emperor Napoleon I) on his expedition to Egypt on 1798–1801. Figure 1.24 shows Napoleon talking the group of scientists who accompanied him on the expedition. Many of their scientific remarks are included in the valuable Description of Egypt series of texts (*Description de l'Égypte*) which was published in 1809 (Figure 1.25).

Figure 1.24 Napoleon Bonaparte talking with scientists on the way to the Egyptian expedition. (By Bingham, peintre anglais.—sabix.revues.org, Public Domain, https://commons.wikimedia.org/w/index.php?curid=38298719.)

Figure 1.25 Frontispiece for the Description de l'Égypte, a work on Egypt commissioned by Napoleon. (By French Government—http://descegy.bibalex.org/Zoom.html?b=1&v=11&p=8&t=undefined, Public Domain, https://commons.wikimedia.org/w/index.php?curid=17723739.)

1.14 *Historical universities offering engineering education*

In this section, a quick preview of the oldest universities in the six continents that are still operating and currently encompass different models of academic engineering degrees is illustrated. It is important to remark that the following list is not a comprehensive one; nevertheless, it is intended to be a glimpse of the progress by which the engineering education evolved and witnessed in world history.

1.14.1 Europe

As already mentioned, European universities had its leading historical role in the world of science, arts, and engineering. The aforementioned analysis considered the ancient and the medieval era. It should also be noted that most of the old universities in Europe founded during that time were mainly devoted to, and still offering, studies in humanities, arts, medicine, and theological studies. However, there are other older universities in Europe which are still in operation and offering engineering degrees. The following is a list of a few of those oldest institutions, representing the premodern era, and ordered by the date of establishment.

Among the older European universities such as Paris, Oxford, and Cambridge, the University of Salamanca is a historic Spanish higher education institution, founded in 1218 in the city of Salamanca, west of Madrid. It is one of the oldest universities in the entire world still in operation. The University is currently offering undergraduate and graduate degrees in a number of engineering fields.

The University of Padua is an Italian university located in the city of Padua, Italy. The University of Padua was founded in 1222 as a school of law and was one of the most prominent universities in early modern Europe. The University of Padua is still in operation, and it has engineering education, undergraduate and graduate programs in industrial engineering, management engineering, information engineering, mechatronics, product innovation, and engineering of materials and nanostructures.

The University of Coimbra is a public university in Lisbon, Portugal, which was established in 1290. It is one of the oldest universities in continuous operation in the world, the oldest and the largest university of Portugal. The university is organized into several faculties granting academic bachelors, masters, and doctorate degrees in nearly all major fields of knowledge, including engineering and technologies.

Within the University of Coimbra, the Faculty of Sciences and Technology has 11 departments offering programs leading to the bachelor and graduate degree programs in design, multimedia, and geospatial information engineering.

Founded in 1592, the University of Dublin is Ireland's oldest operating university. It was modeled after the collegiate universities of Oxford and Cambridge, but unlike these only one college was established; as such, the designations "Trinity College" and "University of Dublin" are usually synonymous. The University of Dublin is one of the ancient universities of Britain and Ireland which is still in operation. Under the school of engineering, the university offers undergraduate and postgraduate programs in civil, structural and environmental engineering, electronic and electrical engineering, and mechanical and manufacturing engineering.

The University of Glasgow is one of Scotland's four ancient universities (Edinburgh, Aberdeen, and St. Andrews). The university was founded in 1451 and is still in operation. A bachelor's of engineering (BEng) and master's of engineering (MEng) are among the main professional degrees awarded by the University of Glasgow.

1.14.2 Asia

In Japan, with its 721 Japanese universities, Ryukoku University (established in 1639) is the oldest. It is a private higher education institution located in the urban setting of the large city of Kyoto. The university offers courses and programs leading to bachelor's degrees in several areas of engineering. Although ranked as the first university in Japan, it was not before (1877) that the University of Tokyo was established.

The University of Santo Tomas is a private university in Manila, Philippines. It was founded in 1611 and is the oldest extant university in the country and in Asia. The university is currently composed of several autonomous faculties, colleges, schools, and institutes, each conferring undergraduate and postgraduate degrees, and the basic education units. In 1907, the Faculty of Engineering was founded. Currently, it offers a bachelor's of science in chemical engineering, civil engineering, electrical engineering, electronics, industrial engineering, mechanical engineering, and communications engineering.

The foundation of Sungkyunkwan University, South Korea dates back to 1398 which makes it the oldest in Asia. However, since its establishment, it was mainly devoted to natural sciences and humanities. A graduate school of mobile systems engineering, sponsored by Samsung Electronics, was established in 2006.

Other Asian universities currently offering engineering curricula were founded back in the 19th century, for instance, the University of Indonesia, Indonesia (1849), and University of Calcutta and University of Mumbai, India (both in 1857).

Extended mostly in Asia and Europe, Russia, with its 379 universities, is listed here. St. Petersburg State University is the oldest Russian university (founded in 1724 by Peter the Great). It is a public higher education institution located in the city of Saint Petersburg, Russian Federation. This university has currently an enrollment of about 25,000 students. The university offers courses and programs leading to bachelor degrees in several areas of engineering studies.

In China, with its 861 Chinese universities, Wuhan University (established in 1893) is a public higher education institution located in Wuhan, People's Republic of China. Wuhan University has currently an enrollment of over 45,000 students. The university offers courses and programs leading to higher education degrees such as bachelor degrees in several areas engineering studies.

1.14.3 North America

Harvard University was founded in 1636, and it was named Harvard College in 1639, chartered in 1650. It is the oldest institution of higher education in the United States. It was officially recognized as a university by the Massachusetts Constitution of 1780. The engineering school within Harvard University's Faculty of Arts and Sciences initially launched in 1847. Presently, it offers undergraduate and graduate degrees in engineering and applied sciences.

The U.S. Congress authorized the creation of the post of chief engineer and the corps of engineers for the army in 1775. The engineers were responsible for building fortifications, surveying terrain, and clearing roads during the independence war and they proved to be so valuable to the revolutionary forces.

Established in 1785, University of New Brunswick (UNB) is a higher education institution which is located in Fredericton town, Canada. UNB has currently an enrollment of around 15,000 students. It offers undergraduate programs in several engineering specializations.

Rensselaer Polytechnic Institute was established in 1824 for the "application of science to the common purposes of life" and is described as the oldest technological university in the English-speaking world. Numerous American colleges or departments of applied sciences were modeled after Rensselaer. During the second half of the 19th century, several universities were established all over the USA and Canada with majors basically in mechanical and civil engineering.

1.14.4 Africa

Even though early engineering evidences could be traced back and contributed to Africa, it was not until the 19th century to witness true engineering education in the continent. Mohamed Aly Pasha (1805–1949), ruler of Egypt, was convinced that modern western education was essential to establish a flourishing and powerful state. Thus, the beginning of modern engineering education in Egypt dates back to 1816 when he established "Madrasat El-Mohandeskhana" (School of Engineering). In 1858, a school for irrigation works at the Barrage and a school for the building were established.

Cairo University, which was established in 1908 as the oldest university in Africa, holds the traces of such schools in its current college of engineering. Mohamed Aly also conducted a series of missionaries for Egyptian students to western schools, mainly to France and some other European countries.

The result was a group of engineers specialized mainly in civil, architecture, mechanical, and manufacturing. Mohamed Mazhar, Aly Mubarak, and many others who were high caliber engineers educated in Europe and participated in modern engineering projects during their times.

Other old institutions in Africa include Fourah Bay College, which is currently a public university in Freetown, Sierra Leone and was originally founded in 1827 as an Anglican missionary school. It is considered the oldest university and the first western-style university built in West Africa. Currently, the college contains a faculty of engineering. University of Cape Town, South Africa was also initiated early in 1829, as a school for boys, which makes it the oldest extant university in Sub-Saharan Africa. The university presently constitutes a faculty of engineering among its affiliated colleges.

There are other old historical universities in Africa, Al-Azhar University, Cairo (established in 975), the University of Al Quaraouiyine (founded in 859 AD) in Morocco, which sometimes considered being the oldest existing, and continually operating educational institution in the world, and the University of Ez-Zitouna in Tunisia (established in 1300). Except for Al-Azhar University which currently incorporates a faculty of engineering (founded in 1961), the other two universities started and continued to be representatives of Islamic religious schools.

1.14.5 Oceania

Established in 1850, the University of Sydney is the oldest institution of its kind in Australia. It is a nonprofit public higher education institution located in the urban setting of the large city of Sydney, New South Wales, Australia. The University of Sydney is officially accredited by the Department of Education and Training, Australia. Currently, it has an enrollment of over 45,000 students. The University of Sydney offers engineering courses and programs leading to bachelor's degrees, master's degrees, and doctorate degrees in several areas of study. These areas include aeronautical engineering, agricultural engineering, architectural engineering, biomedical engineering, chemical engineering, civil and environmental engineering, computer and it engineering, electronic and electrical engineering, general engineering, geological engineering, industrial engineering, mechanical/manufacturing engineering, mining, and metallurgical engineering, and other engineering studies.

1.14.6 Latin America

The National University of San Marcos is a public university in Lima, Peru. It is the first officially established (chartered in 1551) and the longest continuously operating university in the Americas. The University of San Marcos has 60 academic-professional schools, organized into 20 faculties, and 6 academic areas. The university comprises a faculty of engineering which offers degrees in geological, mineral, civil and metallurgical engineering, systems and information engineering, agro-industrial engineering, electronics, electrical and telecommunications engineering, and industrial and textile engineering.

1.15 Epilogue: Engineering is the soul and mind of progress

Throughout this chapter, we took the reader through a journey of time, some thousands of years ago, following evidence of engineering education and application. Presenting landmarks from all over the world, whether these landmarks still exist or only ruins of it is what remains today, it proves that proper engineering principles were applied, and some education of these principles must have been in place, even if schools, colleges, or universities did not exist at that time. Then several contributors to maths and sciences, which are closely related to engineering, at different ages and from different countries were presented. Finally, universities around the globe that included engineering or still offering engineering programs were presented highlighting the different programs offered by these universities.

To conclude, engineering is truly a major contributor to the developing and even existence of mankind. Without engineering and its proper education, man would have not adapted to changes in nature or make use of the resources needed for humanity to develop and prosper.

Bibliography

1794–1804: Revolution and Napoleonic Period (2018) École polytechnique. Available at: https://www.polytechnique.edu/en/revolutionnapoleonicperiod (Accessed: 20 October 2018).

About—St Petersburg University (2018) St Petersburg University. Available at: http://english.spbu.ru/our-university (Accessed: 20 October 2018).

About the Faculty of Engineering and Information Technologies (no date) The University of Sydney. Available at: https://sydney.edu.au/engineering/about.html (Accessed: 20 October 2018).

Al-Azhar University (2018) Wikipedia. Available at: https://en.wikipedia.org/wiki/Al-Azhar_University (Accessed: 20 October 2018).

Annalee Newitz (2013) The Great Library at Alexandria Was Destroyed by Budget Cuts, not Fire, Gizmodo Media Group. Available at: https://io9.gizmodo.com/the-great-library-at-alexandria-was-destroyed-by-budget-1442659066 (Accessed: 11 August 2018).

Anthemius of Tralles (2018) Wikipedia. Available at: https://en.wikipedia.org/wiki/Anthemius_of_Tralles (Accessed: 7 August 2018).

Blake Ehrlich, Richard R. Ring, and John Foot (2018) Rome, Encyclopædia Britannica, Inc. Available at: https://www.britannica.com/place/Rome (Accessed: 11 August 2018).
Callicrates (2018) Wikipedia. Available at: https://en.wikipedia.org/wiki/Callicrates (Accessed: 7 August 2018).
Carole Escoffey (2012) Ancient Alexandria. Available at: https://www.bibalex.org/Attachments/Publications/Files/2013032015003765555_AncientAlexandria.pdf (Accessed: 7 August 2018).
Civilization: Ancient Mesopotamia (2018) TimeMaps Ltd. Available at: https://www.timemaps.com/civilizations/ancient-mesopotamia/ (Accessed: 7 August 2018).
Dominique Sourdel (2018) Al-Ma'mūn, Encyclopædia Britannica, Inc. Available at: https://www.britannica.com/biography/al-Mamun#ref248577 (Accessed: 7 August 2018).
History com Editors (2009) Morse Code & the Telegraph, A&E Television Networks. Available at: https://www.history.com/topics/inventions/telegraph (Accessed: 20 October 2018).
Engineering Education History (2018) Faculty of Engineering—Cairo University. Available at: http://eng.cu.edu.eg/en/engineering-education-history/ (Accessed: 20 October 2018).
Fourah Bay College (2018) Wikipedia. Available at: https://en.wikipedia.org/wiki/Fourah_Bay_College (Accessed: 20 October 2018).
Francis Bacon (2018) Wikipedia. Available at: https://en.wikipedia.org/wiki/Francis_Bacon (Accessed: 20 October 2018).
Galileo Galilei (2018) Wikipedia. Available at: https://en.wikipedia.org/wiki/Galileo_Galilei (Accessed: 20 October 2018).
Harvard University (2018) Wikipedia. Available at: https://en.wikipedia.org/wiki/Harvard_University (Accessed: 20 October 2018).
Hero of Alexandria (2018) Wikipedia. Available at: https://en.wikipedia.org/wiki/Hero_of_Alexandria (Accessed: 7 August 2018).
High School Engineering/Engineering in Medieval and Renaissance Europe (2015) Wikibooks. Available at: https://en.wikibooks.org/wiki/High_School_Engineering/Engineering_in_Medieval_and_Renaissance_Europe (Accessed: 11 August 2018).
High School Engineering/The Industrial Revolution (2015) Wikibooks. Available at: https://en.wikibooks.org/wiki/High_School_Engineering/The_Industrial_Revolution (Accessed: 20 October 2018).
Higher Technical School of Industrial Engineering of Béjar (2016) Universidad de Salamanca. Available at: http://www.usal.es/escuela-tecnica-superior-de-ingenieria-industrial-de-bejar (Accessed: 20 October 2018).
Historical Review (no date) Universidad Nacional Mayor de San Marcos. Available at: http://www.unmsm.edu.pe/home/inicio/historia (Accessed: 20 October 2018).
History com Staff (2010) Great Wall of China, A+E Networks. Available at: https://www.history.com/topics/great-wall-of-china (Accessed: 11 August 2018).
Johannes Gutenberg (2018) Wikipedia. Available at: https://en.wikipedia.org/wiki/Johannes_Gutenberg (Accessed: 20 October 2018).
Johannes Kepler (2018) Wikipedia. Available at: https://en.wikipedia.org/wiki/Johannes_Kepler (Accessed: 20 October 2018).

Joshua J. Mark (2016a) Great Pyramid of Giza, Ancient History Encyclopedia. Available at: https://www.ancient.eu/Great_Pyramid_of_Giza/ (Accessed: 7 August 2018).

Joshua J. Mark (2016b) Imhotep, Ancient History Encyclopedia. Available at: https://www.ancient.eu/imhotep/ (Accessed: 7 August 2018).

Joshua J. Mark (2018) Mesopotamia, Ancient History Encyclopedia. Available at: https://www.ancient.eu/Mesopotamia/ (Accessed: 30 July 2018).

Leonardo da Vinci (2018) Wikipedia. Available at: https://en.wikipedia.org/wiki/Leonardo_da_Vinci (Accessed: 20 October 2018).

List of Early Modern Universities in Europe (2018) Wikipedia. Available at: https://en.wikipedia.org/wiki/List_of_early_modern_universities_in_Europe (Accessed: 20 October 2018).

List of Oldest Universities in Continuous Operation (2018) Wikipedia. Available at: https://en.wikipedia.org/wiki/List_of_oldest_universities_in_continuous_operation (Accessed: 20 October 2018).

List of Universities in Germany (2018) Wikipedia. Available at: https://en.wikipedia.org/wiki/List_of_universities_in_Germany#Universities_by_years_of_existence (Accessed: 11 August 2018).

Mark Cartwright (2017) Paper in Ancient China, Ancient History Encyclopedia. Available at: https://www.ancient.eu/article/1120/paper-in-ancient-china/ (Accessed: 15 August 2018).

Menso Folkerts et al. (2018) Mathematics in the Islamic world (8th–15th century), Encyclopædia Britannica, Inc. Available at: https://www.britannica.com/science/mathematics/Mathematics-in-the-Islamic-world-8th-15th-century#ref536168 (Accessed: 11 August 2018).

Michael Deakin (2018) Hypatia, Encyclopædia Britannica, Inc. Available at: https://www.britannica.com/biography/Hypatia (Accessed: 11 August 2018).

Mimar Sinan (2018) Wikipedia. Available at: https://en.wikipedia.org/wiki/Mimar_Sinan (Accessed: 20 October 2018).

New World Encyclopedia contributors (2016) University of Paris, New World Encyclopedia. Available at: http://www.newworldencyclopedia.org/p/index.php?title=University_of_Paris&oldid=1000964 (Accessed: 11 August 2018).

Oldest Universities in Australia (2018a) uniRank. Available at: https://www.4icu.org/au/oldest/ (Accessed: 20 October 2018).

Oldest Universities in China (2018b) uniRank. Available at: https://www.4icu.org/cn/oldest/ (Accessed: 20 October 2018).

Oldest Universities in Russia (2018c) uniRank. Available at: https://www.4icu.org/ru/oldest/ (Accessed: 20 October 2018).

Our history—University of Bologna (no date) Università di Bologna. Available at: https://www.unibo.it/en/university/who-we-are/our-history (Accessed: 11 August 2018).

Persepolis (2018) Wikipedia. Available at: https://en.wikipedia.org/wiki/Persepolis (Accessed: 20 October 2018).

PhD Programmes (2018) University of Padova. Available at: https://www.unipd.it/en/phd-programmes (Accessed: 20 October 2018).

Ralph J. Smith (2017) Engineering, Encyclopædia Britannica Inc. Available at: https://www.britannica.com/technology/engineering (Accessed: 30 July 2018).

Rensselaer Polytechnic Institute (2018) Wikipedia. Available at: https://en.wikipedia.org/wiki/Rensselaer_Polytechnic_Institute (Accessed: 20 October 2018).

Richard S. Westfall (2018) Sir Isaac Newton, Encyclopædia Britannica, Inc. Available at: https://www.britannica.com/biography/Isaac-Newton (Accessed: 11 August 2018).
Samuel Brown (Engineer) (2018) Wikipedia. Available at: https://en.wikipedia.org/wiki/Samuel_Brown_(engineer) (Accessed: 20 October 2018).
School of Engineering (2018) Trinity College Dublin, The University of Dublin. Available at: http://www.tcd.ie/Engineering/ (Accessed: 20 October 2018).
Sultan Ahmed Mosque (2018) Wikipedia. Available at: https://en.wikipedia.org/wiki/Sultan_Ahmed_Mosque (Accessed: 20 October 2018).
Sungkyunkwan University (2018) Wikipedia. Available at: https://en.wikipedia.org/wiki/Sungkyunkwan_University#Facilities (Accessed: 20 October 2018).
Taj Mahal (2017) Britannica Academic. Available at: https://academic.eb.com/levels/collegiate/article/Taj-Mahal/70996 (Accessed: 15 August 2018).
The Editors of Encyclopaedia Britannica (2017) Apollonius of Perga, Encyclopaedia Britannica, Inc. Available at: https://www.britannica.com/biography/Apollonius-of-Perga (Accessed: 11 August 2018).
The Editors of Encyclopaedia Britannica (2018) Forbidden City, Encyclopædia Britannica, Inc. Available at: https://www.britannica.com/topic/Forbidden-City (Accessed: 15 August 2018).
The Wealth of Nations (2018) Wikipedia. Available at: https://en.wikipedia.org/wiki/The_Wealth_of_Nations (Accessed: 20 October 2018).
Theaetetus (Mathematician) (2018) Wikipedia. Available at: https://en.wikipedia.org/wiki/Theaetetus_(mathematician) (Accessed: 7 August 2018).
Today in History—June 16 (2018) Library of Congress. Available at: https://www.loc.gov/item/today-in-history/june-16/ (Accessed: 20 October 2018).
Top Universities in Japan (2018) uniRank. Available at: https://www.4icu.org/jp/ (Accessed: 20 October 2018).
Tycho Brahe (2018) Wikipedia. Available at: https://en.wikipedia.org/wiki/Tycho_Brahe (Accessed: 20 October 2018).
Undergraduate Studies (2018) University of Glasgow. Available at: https://www.gla.ac.uk/undergraduate/choosingyourdegree/ (Accessed: 20 October 2018).
University of Calcutta (2018) Wikipedia. Available at: https://en.wikipedia.org/wiki/University_of_Calcutta (Accessed: 20 October 2018).
University of Cape Town (2018) Wikipedia. Available at: https://en.wikipedia.org/wiki/University_of_Cape_Town#Rankings (Accessed: 20 October 2018).
University of Dublin (2018) Wikipedia. Available at: https://en.wikipedia.org/wiki/University_of_Dublin (Accessed: 20 October 2018).
University of Glasgow (2018) Wikipedia. Available at: https://en.wikipedia.org/wiki/University_of_Glasgow (Accessed: 20 October 2018).
University of Indonesia (2018) Wikipedia. Available at: https://en.wikipedia.org/wiki/University_of_Indonesia (Accessed: 20 October 2018).
University of Mumbai (2018) Wikipedia. Available at: https://en.wikipedia.org/wiki/University_of_Mumbai#Academia (Accessed: 20 October 2018).
University of New Brunswick (2018) uniRank. Available at: https://www.4icu.org/reviews/596.htm (Accessed: 20 October 2018).
University of Santo Tomas (2018) Wikipedia. Available at: https://en.wikipedia.org/wiki/University_of_Santo_Tomas#Faculties (Accessed: 20 October 2018).
University of Tokyo (2018) Wikipedia. Available at: https://en.wikipedia.org/wiki/University_of_Tokyo (Accessed: 20 October 2018).

Vicente Rodriguez (2017) Andalusia, Encyclopædia Britannica, Inc. Available at: https://www.britannica.com/place/Andalusia-region-Spain (Accessed: 11 August 2018).

Welcome to FCTUC (2018) University of Coimbra. Available at: http://www.uc.pt/en/fctuc (Accessed: 20 October 2018).

chapter two

Strategies-challenges of engineering education

Osman Taylan and Abdullah Bafail
King Abdulaziz University

Contents

2.1 Introduction

Currently, engineering education and its assessment face various challenges in meeting the diverse needs of students, growing and changing demands of global industrial environment. There is a rising need for high-skill staff across several industries in the globe for intelligent systems-related fields (Artificial Intelligence (AI)), the robotics industry, manufacturing industry, software systems, oil and gas industry, nuclear power plants, and in manufacturing fields where certain intelligent appliances are produced, and in traffic control systems. The demand for high-skill multitalented engineers is growing, and parallel to this, the employment opportunities will continue to grow in near future. Therefore, the challenges are becoming quite different for employing, retaining, and supporting enough students to graduate as engineers who are prepared to fill these positions and who will value and pursue continuously learn or keep on learning during their life. In engineering education, the most common instructive (pedagogical) approaches in the framework of learning are based on students' competency which includes experiential learning, project-based learning, problem-based learning, and team-based learning. Because the number and types of competencies required for the aspirants to be competitive in the job market are changing very fast due to the market demands and changing work environment. The competency-based student learning (CBSL) can become accustomed easily to these changes and the requirements for the engineers of future. Froyd et al. (2012) claim that there is a shift toward a competency-based learning in the Europe and also in many universities in United States, Asia, Australia, and South America (Felder et al., 2011). Once the number and types of competencies are determined, further competencies can be fostered in the engineering curriculum by breaking them into discrete or subparts that can be evaluated and assessed separately. In order to prepare students for the workplace of future, CBSL can add benefits as an effective and continuous learning approach to prepare the students for the future. CBSL encourages a strong emphasis on self-directed continuous learning beyond the classroom; it is a lifelong learning approach and has been identified as one of the key characteristics needed for success in industry of future. Spelt et al. (2015) state that self-learning can be effective to improve not only the knowledge-based capabilities needed for the engineers but also the professional skills that are equally important for engineers in the current global industrial world. Hence, a question can be raised about "the mission of universities to contribute and progress the engineering society by pursing the creation, preservation and distribution of knowledge." Knowledge is created and achieved by means of research and development (R&D) tasks. The teaching–learning activities are used for the preservation and dissemination of knowledge, which is for promoting and supplying it to the professionals of future. Hence, the generation of patents, prototypes of products, and publications of papers are all carried out by learning more and better not only for knowledge dissemination but also for technology transfer and/or innovation actions. Engineering students should be placed at the central point of the analyses carried out in relation with universities, their learning level, and assessment methods. It is clear that research activities at universities, without the participation of students taking part in or benefiting from them, are out of focus and against knowledge dissemination. Hence, engineering education must be focused on the full development of the student personality and provide them with the intellectual background, and social environment to help them achieving their dreams, and become a successful person in their professional life. In addition, in UN General Assembly (2015), it was said that education should be available and equally accessible to all on their basis of merit. The route and root of future engineering education requires the best trained engineers for further developments to mentor the technological advances that are reshaping the life of human being by including the technological revolution, the planet-wide communication grids, the advent of nanotechnology, and the emergence of AI, among others (Díaz Lantada et al., 2016). There is an increasing need to produce competent engineers to meet the employment gap for the demand of industry. Engineering education institutions face the challenge of producing engineers prepared to fill many positions as well as ensuring that their graduates possess the necessary competencies to succeed in the workplace. Engineering education for future aims to meet several challenges and to develop strategies for systematically promoting the learning level. If the *engineers of the future* are to be key actors in solving present-day challenges of mankind, most talented and motivated engineers will be needed, regardless of their social background and economic status. In fact, modern engineering programs merge theoretical knowledge of basic disciplines with science in-depth and technology for more applied actions not only for technical skills but also for fundamental professional outcomes, so as to educate successful engineering professionals. The applied activities in engineering education are the practical in laboratories with state-of-the-art technologies, project-based learning activities, research projects, visiting industrial sites, professional practices, and the final degree theses. Such a combination between theoretical and practical teaching–learning strategies helps to configure interesting curricula for building well-trained professionals, dedicated teaching staff for continuous training, as well as well-equipped laboratories and research centers with advanced technologies. All these activities make engineering education more complex and expensive than the other disciplines. Engineering education requires carrying out systematic studies and is based on process re-engineering methodologies aimed at continuous improvement, which has been previously applied for the promotion of professional skills (Díaz Lantada et al., 2013a), for the improvement of project-based teaching–learning activities (Díaz Lantada et al., 2013b), and for the overall enhancement of the teaching–learning process (Munoz-Guijosa et al., 2009).

2.2 Strategies on instructors teaching methodologies in engineering education

In fact, students are the central element of the teaching and learning process; on the other hand, instructors' teaching methodology plays a central role in engineering education. In order to let *engineering education for the future* attract closer attention to changes, instructors' knowledge, abilities attitudes play a very significant role. Pappano (2012) and McAuley et al. (2010) suggest that it is important to highlight new teaching–learning paradigms as well as the open-access publication outlines in engineering research, which can re-shape engineering education and helping to promote the concept of *engineering education for the future.* The outcomes of collaboration between instructors, students, families, and student associations for the advancement of *engineering education for the future* must be taken into consideration. For instance, university and industry cooperation have proved to be helpful for continuously enhancing the quality of products, for the effectiveness of industrial processes, and for improving the functionalities of new devices. Relations between university and industry are significantly beneficial for the teaching–learning process in engineering education. Because, it helps to reestablish the syllabi and the topics covered, so as to sustain the speed of changing knowledge-based industries, thus helping to prepare students more for their future responsibilities, and to enhance both their technical and professional talents and competency.

For improving the teaching–learning processes, there are several central facilities, information and communication technologies, and supportive administrative staff who are adding great value for avoiding instructors to be unnecessarily devoted to bureaucratic procedures, instead of devoting their time to teaching, research, and strategic planning. In order to avoid wasting time and increasing the value adding activities; industrial partners, professional associations, and alumni may play important role. Academic staff who are overwhelmed with bureaucratic duties needs additional stimuli for continuously upgrading the teaching–learning methodologies and compromising with lifelong learning, including their participation in research and innovation activities, as well as visiting other relevant research centers and industries for periods of year. On the other hand, employing teaching assistants may be a good strategy for reducing the load of academic staff, and/or additional helps can be provided from central facilities and administration staff, for enabling academic staff to concentrate on teaching and research activities. This will help academics to interact also with the industrial environment by means of joint innovation projects.

2.3 Learning strategies and student competency

CBSL is able to encounter the growing demands of industry for competent engineers by confirming that graduates have grasped the necessary skills to be used in industry. Díaz Lantada et al. (2016) state that the evaluation of student advancement is based on if students are able to show their competencies that are precise and measurable objectives clearly communicated to them. Under the CBSL framework, mastery of competencies includes the ability to apply knowledge in practical real-life problems. CBSL is an outcome-based, student-centered form of education where student progress is advanced by working on mastering the necessary prerequisite content and skills. CBSL, apart from the other instructional approaches, is the modification in focus to behavioral outcomes of the graduates. This method mainly aims for lifelong learning and is away from setting time limits during which a certain amount of knowledge should be gained and learned. The approach allows students to progress their learning level at their own pace, and establish a continuous learning environment for achieving the best level of learning. This approach is also appropriate and similar to the philosophy of lifelong learning. Hence, the approach can revise the objective from increasing the amount of information that can be taught in a semester to ensuring that students' master outcomes before moving on to the next level (Henri et al., 2017). In the CBSL framework, instructional materials, course-learning objectives, and even the exam questions and case-study questions can be related to specific predefined outcomes that are clearly communicated to students at the beginning of any semester. CBSL can be considered a subset of advanced level (mastery) learning. Advanced level learning (ALL) is called deep learning that focuses on achieving the highest competency level before graduation, and learning subsequent information (Tyler, 2013); it is a deep-level learning that students can synthesis and evaluate the results and findings of a real-life problem. ALL has been widely studied as a strategy for improving student outcomes (SO; Bloom, 1984). CBSL is distinguished from other ALL approaches; it has very stringent criteria to constitute the scope of mastery learning level. The approach emphasizes assessing students learning level on measurable competencies. Hence, it means to say that learning skills in CBSL are based not only on theoretical or conceptual understanding of subject matters but also on the soft abilities (skills) to apply the knowledge gained in practical settings (Henri et al., 2017). CBSL can be defined with other terminologies including student-centered learning, self-paced learning, student/self-directed learning, individualized or personalized instruction, outcome-based learning, performance-based learning, and standard-based learning in proficiency-based education (Roe, 2015).

The primary purpose of CBSL is to improve SO regardless of the field of graduation. Henri et al. (2017) claim that to improve student learning achievement, desirable skill sets are determined and broken into subcompetencies that build on each other. In this context, house of quality (HoQ) can help to determine the desirable skills and competencies. CBSL can achieve both goals by individualizing learning, promoting autonomy, encouraging continuous or lifelong learning, and inspiring students to take charge of their own education. The focus of engineering education has shifted from input- to output-based teaching–learning graduation systems, where students master learning outcomes are predetermined before they have moved to more advanced content so that they can see the complete picture of their advancement from the early beginning. Several challenges exist recruiting and retaining enough students to educate, teach, and train in today's world for such diverse needs and graduate them as engineers who are prepared to fill the positions and will value and pursue continuous or lifelong learning. CBSL has the potential to progress engineering graduates' writing and communication skills very well, consider their ethical and social responsibilities, and comprehend and manage the complex engineering systems within a framework of sustainable development and make them ready and/or prepare them to live and work as global citizens. These practices in engineering education need to be recognized in the beginning and used to establish a bridge filling the gap between student education and industry needs.

There are several reasons why CBSL is an effective educational policy. According to Spelt et al. (2015), CBSL is an efficient interdisciplinary field, because it is a student-centered self-learning approach. It leads students to be more self-sufficient (autonomous) and learn how to upgrade themselves from different sources. O'Reilly (2014) states that autonomy is a major factor in student achievement with research indicating that when students have a higher sense of perceived control of their own education, they tend to perform learning better. More autonomous students are likely to be better equipped to achieve the necessary competencies in their fields or disciplines and are better at information and knowledge integration, synthesis, and analysis the findings of problems solved. It is recommended by several faculty using instructional methods that promote student-centered teaching and learning to achieve the necessary competencies. Re-structuring the learning environment for achieving all engineering students in a manner that promotes autonomous behaviors, self-esteem, and self-reliance; competency-based autonomous learning should be one of the outcomes of engineering education. CBSL can promote the autonomous learning and affects the performance of students directly and indirectly through motivation and positive attitudes. Aeschlimann et al. (2016) and Hall and Webb (2014) claim that by increasing student autonomy, and motivation, a positive attitude can be generated toward learning.

2.4 *Learning assessment strategies and structure*

In CBSL, instructors have to obviously communicate with students about the competencies and it has to be taught and the assessment criteria have to be introduced. As a result, students will have a clear roadmap for successfully completing courses. Hence, when students are aware of the learning objectives and goals from the beginning of semester, and if instructors determine the assessment criteria based on pre-established competencies, there will be less student confusion or dissatisfaction with the courses. Gharaibeh et al. (2013) stated that although CBSL has many benefits, there are also documented disadvantages and challenges for student learning. CBSL is an effective approach in teaching student professional skills such as communication, prioritization, working under pressure, and involving in teamwork activities for solution of problems. As it was reported by Roe (2015), engineering courses are weak and rarely help engineering students to develop non-content-based skills, which are then learned only incidentally or accidently. However, professional skills (soft and technical skills) and contextual competencies are essential in the workforce and are difficult to define and break into units that can be taught and assessed during work life. The incompetence discretizing professional skills along with their integration into other curricular aspects of student learning make it challenging to teach students using competency-based learning. Engineering students who lack the key requirements of appropriate competence will feel less confident in their work life than in other students (Ro et al., 2015). On the other hand, De Los Rios-Carmenado et al. (2015) claim that technical engineering competencies alone are not sufficient in the current work environment, which means that the importance of developing the appropriate soft and technical skills and attitudes needed for the today's organization should be emphasized clearly and introduced to engineering students (Woodrow et al., 2013). Engineering graduates are required nowadays to be proficient in their content areas, and also should be effective communicators, negotiators who can convince the others, utilize their knowledge, and use it for the benefit of society and mankind.

One limitation of CBSL approach is that it can limit the capability of students to grasp the big picture of learning tools and learning criteria. Therefore, students may not have a strong chance to grasp each and every piece of knowledge to synthesize and analyze the content of a work. In other words, Walther et al. (2011) claim that CBSL may divide the knowledge into pieces which will make it difficult for students to pick, merge, and learn, and integrate to establish a picture for the judgment of findings. However, knowledge can be addressed by assessment tools including integration of competencies, holding in comprehensive exams, evaluating end of year projects/assignments that require understanding the mastery of not only each competency but also the relationships among discrete tools (Henri et al., 2017). One way to present the limitation is to explicitly link competencies to a larger framework. For instance, ABET (2016) requires that engineering programs link student educational outcomes to predetermined course-learning objectives. This can help students link each competency or skill to a larger framework and establish a picture, and set relations between the learning materials and the SO. There are several challenges that educators encounter when transforming an engineering program from a traditional instructional format to a competency-based framework. An instructor needs to invest more time and effort into making the transition of instructional method changes, and the transition to CBSL which can be difficult for instructors and students who are used to studying and teaching in traditional classrooms and using instructional settings (Rikakis et al., 2013). For instructors, it is considered time intensive to break materials and skills into discrete competencies offering several assessment tools that build upon each other and construct objective assessments for those competencies. As stated by Evans et al. (2015), the student-centered learning focusing on CBSL gives students more autonomy and responsibility, which might be usually a difficult alteration in the beginning. Similar to any change in any field, to establish the CBSL environment, there might be illogical demands, resistance, and challenges. However, once instructors and students have successfully managed the learning curves and witnesses the benefits of self-learning experience and autonomy during restructured ways of spent their time, the organization and results of CBSL will be more efficient use of it (Di Trapani & Clarke, 2012). Professional and content-based competencies are aimed in engineering education.

2.5 Engineering program assessment approaches

Assessment is an important but difficult task in the whole teaching and learning process which has a strong influence on students' learning and learning outcomes (Ma & Zhou, 2000). Educational development has shifted from being teacher-centered learning to student-centered learning, recently. It usually takes place in the form of problem-based learning, project-based learning, and lab-based learning. Criterion-based assessment can be used for the evaluation of students' outcomes due to the nature of student-centered learning. This assessment approach leads the attention of students to their performance with respect to the tasks they undertake and give priority to their competency. However, in the current practice of student-centered learning environment, the instructors in charge of teaching are merely liable for determining the assessment criteria and their corresponding weights. This assessment approach may diminish the students' autonomy in the whole learning process and lower their learning quality level. Generating the basic set of assessment criteria with reference to the course objectives and the industry standards, fuzzy set approach can be employed for the assessment of student learning. The criteria agreed on are then used to evaluate the students' outcomes on a fuzzy grading scale. Quality control charts can be used to scale the level of learning and assess the quality of student learning. Fuzzy set theory is widely used for the evaluation and assessment of problems. The criteria *learning* and its term set *very poor, poor, average, good, very good, and excellent* are uncertain and vague fuzzy sets used for the explanation of student learning assessment level. A fuzzy set in student learning assessment level is a class of sets with a variety of membership degrees. A membership function (MF) can be used for each term to grade the membership degrees and associate them with a fuzzy set. The nature of student learning is not precise; currently, numerical values are used to assess the level of learning. However, the precise assessments are not possible and very suitable in this environment. Although the membership grades determined through membership values of fuzzy terms do not convey any absolute significance, they can instead define an interval for learning level which is more meaningful. Fuzzy sets are context-dependent and can be subjectively assessed. Fuzzy set theory can be employed in the assessment and evaluation of student learning at course level as well. As it is well known, the most commonly assessment tools used are *take-home exams, comprehensive, or lab exams; teamwork, such as projects and presentations,* are widely used in student learning-level assessment. Similarly, visual, digital/electronic materials, oral presentations, and paper content are electronic portfolios used to evaluate not only knowledge-based competencies but also professional competencies (soft skills) as well. Effective communication, well organization, and teamwork are tools used by some instructors to increase the motivation and autonomy in learning, while giving students the opportunity to demonstrate their professional competencies with writing an article and/or making an effective presentation to increase a student's communication skill (Badilla Quintana et al., 2014). De Los Rios-Carmenado et al. (2015) used 360-degree assessment strategies to improve the acquisition of content-based learning and improve soft skills such as communication, critical thinking, writing, and time and task management and promoting the attitudes of students.

Soft skills aim students to have certain competencies including analytical thinking, verbal and written communication, and leadership skills. One reason soft skills are so revered is that they help facilitate human connections. Hence, soft skills are key to building relationships, gaining visibility, and creating more opportunities for advancement. These skills are *values, having strong beliefs, independent thinking, teamwork skills, mindfulness of others rights, and respect to others.* Soft skills can be detailed as follows;

For instance, *value* is not a simple soft task to teach students. It includes principles and standards of behaviors; it is the judgment of what is important in the life for a person. It can also be defined as *merit*, helpfulness and the regard that something is held to deserve, the importance, worth, or usefulness of something. Additional values could be count for students are; having chance to study art, painting, sports, and music.

The *beliefs* mean holding something as an opinion in the mind; a standard of life, a way of servicing people more respectfully and just in time, and a way of thinking to carry out duties properly. It also means learning how to say "no" for something out of the frame of your beliefs. This does not mean that imposing your ideas to others. It is to be an open-minded student who can establish high standard for himself and for his friends and open to discuss the ideas brought by team members.

Independent thinking helps educational institutes and young people be better, both in the universities and beyond. Because students are not regarded as a source of data, their success and skills cannot be properly measured by numerical data; therefore, they deserve to get chances to show their abilities and skills in the school and work life, obedience is not engagement, silence is not respect, learning is more than memorizing, and there is always another way to perform the jobs. Mastering independent thinking empowers a student in order for him/her to develop skills to solve a critical problem. Independent thinkers are good team leaders because they consider different backgrounds and ideas of their team and help finding faster solutions to problems. Independent thinking is to encourage students to ask questions and then find the solutions on their own. Encouraging them to carry out experiments, then explain their results and defend the conclusions, will make them more successful. Students must know that independent thinking skills will help them to be more efficient and empower them to do better in the classrooms and outside. If they fail the first time, we must teach them to try, and try again. These lifelong skills will help them to be successful.

Teamwork involves building relationships with other students and working closely with them using a number of important skills and habits such as working cooperatively, contributing to group with ideas, suggestions, and effort, establishing two-way communication with them, developing sense of responsibility, healthy respect for different opinions, and individual preferences, and ability to participate in group decision-making. Understanding the value of teamwork and becoming an effective team member are important soft skills to develop leadership skills. Having students gain experiences through which they learn to be sure of themselves and others is an important factor in the development of a productive teamwork mentality. The teamwork will encourage participants to explore different aspects of a problem, as well as the ways that they can become a top-notch team performer. By this way, the team participants will be given concepts of what makes up a team, and details about which factors should be taken into consideration for being a successful team and team member.

Mindfulness is related to self-control to focus on increasing the sense of gratitude through the recollection of the kindness that one receives from others. It is also to be aware of surroundings, to understand "what someone is doing." Although mindfulness is nowadays popularly perceived as the practice of being aware of one's surroundings, actions, inner state of mind, and/or sensory experiences, and mindfulness to involve reflection on certain existential facts of life. There are many similarities between the concepts of mindfulness that a student who regularly exercises his/her mindful-muscle in the process of mastered self-control. It can be explained by other concepts too, such as kindness to others, and staying positive to all. Mindfulness also reflects the rays of an inner sun which only seeks to bring light of happiness and growth all whom it touches. Thus, mindfulness cannot exist without a moral compass.

Being respectful to others, if students want to improve their relationships, is necessary and important in business and/or personal relations with others. For instance, respect is an indispensable part of human life. It is an unspoken way of communication which builds unshaken and strong relations among people. When one shows respect to another, then it means that he/she has some values for the others that might be important advices and suggestions. Hence, students must have the following soft skills underlined for the basis of respect: *listening other* which can be one of the hardest skills to have. "Encouragement of people," "congratulating others," "being helpful," and "saying thank you to others" are the important soft skills that a student is advised to have.

The electronic or online assessment tools can be built as a way to evaluate knowledge and learning level of students before beginning a new unit or subject and determine the extent of learning at the end of the electronic module application in the form of a post-test. Online student learning assessment can be a time-efficient and sustainable system for continuous monitoring of student progress in teaching and learning environment (Al-Thani et al., 2014). As it is suggested by Roe (2015), the assessment tools can be in the form of electronic quizzes taking very short time, and exams that students take as soon as they finish a subject in a topic covered or a section of a module. Student's performance on these evaluation tools may indicate if they have built sufficient mastery knowledge of the content to move to the next subject or a new module. Furthermore, these online testing systems can provide accommodations for computer adaptive evaluation of student knowledge and provide students with automated feedback based on their learning level (Hsu & Ho, 2012); hence, these tests provide students and instructors the weak points of teaching and learning process. The assessment strategies listed above usually focus on technical competencies and skills rather than soft (professional) competencies and skills.

On the other hand, there are indirect assessment tools carried out by surveys and questionnaires to determine students' level and their improvement in building the technical and soft skills. Surveys and questionnaires can play a central role in the CBSL assessment. Surveys can also find out and determine the expectations of students during teaching and assessment. In this context, Taskinen et al. (2015) employed a competency scale that was psychometrically evaluated and validated to capture content-based competencies in a course. On the other hand, surveys and questionnaires can be used either as written feedback or, more efficiently, as an online tool to use as an indicator of student learning level and their expectations from engineering program to achieve high quality of learning level. Management of modern teaching–learning systems requires interactive instructional methods that can provide instant feedback, and direct students to relevant additional content-based materials and allows them to advance to more difficult contents only after competence is shown (Henri et al., 2017). In engineering education, the competency-based learning includes experiential learning, project-based learning, problem-based learning, and team-based learning. Experiential learning is broadly defined as a purpose that instructors employs students with indirect experience and focused reflections to increase their knowledge, develops their soft and technical skills, and clarifies values that they may need in the future. Grodzicki and Madigan (2011) determined that experiential learning activities are a key to successfully preparing students for the companies, which have become more widespread in the last decade for improving student learning, fostering deep learning, and understanding of meta-knowledge. Bogue et al. (2013) suggested that a successful engineering program must contain hands-on activities, interdisciplinary activities, and problem-solving tools to improve teamwork capabilities of students to prepare them for obtaining business oriented soft and technical skills, and obtain real-life engineering experiences. In this context, experienced students can share their competencies; hence, alumni and faculty can be called to share their experiences. There are several forms and techniques of AI such as virtual reality, expert systems, and augmented reality systems which can serve as supportive tools to aid students in learning for more positive outcomes for the future.

Project-based and problem-based learning are well-known and widely used educational approaches and effective in revealing students' mastery to achieve higher learning level. Project-based learning can be the occasion of improved student learning outcomes in the case that the project undertaken is a comprehensive engineering project. Instructors and researchers may allow students to participate in engineering projects to improve their competency by applying some certain tools of engineering for critical thinking and writing their reports as an assessment tool for problem-based advanced learning. Problem-based learning promotes achieving high level of learning by thinking strategies and domain knowledge and leading students through the experience of solving open-ended problems.

Similarly, team-based learning is an instructional strategy to integrate the knowledge of students by focusing on utilization of collaboration and exchanging knowledge among them. The reality is that students learn a lot from each other during the team-based learning; however, domination of a student should be prohibited during the teamwork. In this context, one of the challenges that may involve in implementation of team-based learning is social loafing. The practice of team members relying on the work of others rather than contributing a reasonable effort to the endeavor is called loafing. Zou and Ko (2012) revealed that, in teams, social loafing widely existed, but students tended not to report it or hold the loafers accountable. This can be prohibited by holding a small interview exam with the team members during submission of projects. By this way, the instructor can determine how much a student has contributed to the teamwork. The instructor can divide students into small groups and assign each group a task requiring the utilization of relevant knowledge and competencies. The instructor in this case acts as an expert advisor, a coach, or a facilitator providing feedback to the teams throughout the task. Team-based learning can be combined with project-based learning by assigning student teams different projects aiming to improve overall learning outcomes (Henri et al., 2017). Collaborative learning is a similar approach and seems to mitigate some of the limitations of team-based learning (Canaleta et al., 2014).

2.6 *Professional competencies in engineering education*

Global engineering education for the future can be built with CBSL which will be effective in enabling engineers proficient in technical knowledge skills, and will have the soft skills ready to compete in global world by possessing the ability to successfully work with diverse groups of people in diverse contexts. Although knowledge-based competencies change in different engineering disciplines, some professional competencies are required for all engineers when they receive their undergraduate degrees. These professional competencies and some other skills are essential for students to be competitive and proficient engineers in the business organization. The following outcomes can prepare engineering graduates for workplaces in a dynamic, interdisciplinary, and universal environment. Table 2.1 presents the learning outcomes and its associated skills.

Similarly, Ro et al. (2015) underlined that the robust competencies are contextual competencies and effective communication that are mainly soft skills and related to:

- knowledge of social, cultural, political, economic, and environmental contexts that may affect a solution to an engineering problem,
- knowledge between technological solutions and the context in which they operate,
- ability to use knowledge from different cultures, social values, political systems, economic conditions, and environmental issues to guide engineering solutions (Henri et al., 2017).

Table 2.1 Student learning outcomes and associated skills

Learning outcomes (technical and soft skills)	Associated skills
Critical and creative thinking	Inquiring and analyzing knowledge, solving problems, creativity, depth, and breadth of understanding.
Literacy and contextual competencies	Information, quantitative, technological, and visual literacy.
Global understanding	Global understanding skills, sense of historical development, intercultural competence.
Emotional intelligence and leadership skills	Self-awareness, self-regulation, motivation, empathy, social skills, leading skills.
Complex problem-solving capability	Defining the problem, generating alternatives, evaluating, and selecting alternatives, implementing solutions.
Multidisciplinary perspective and teamwork skills	Communication, conflict management, listening, reliability, respectfulness
Judgment and decision-making capability	Participation, deliberation, diversity of thought, constructive challenge, debating
Cognitive flexibility and communication skills	Open-minded perspectives, willingness to risk mistakes, multiple ways to solve problems, engagement in learning, discovery, and problem-solving with innovative creativity; Oral, written, and integrative communication, reading comprehension.
Interdisciplinary skills	Recognition, in-attentional blindness, embrace divergent thinking, and harnessing the potential of transferring their learning.
Professional and ethical behavior and *trustworthiness*	Ethical reasoning, professional leadership, teamwork, personal organization, and time management.

Although importance of the competencies has been emphasized so far, still, there is a gap in the literature on clear persistent strategies for assisting students to gain these competencies and appropriate ways for assessing them. Actually, in many engineering programs, these competencies are accidently or incidentally attained as a result of the college experience.

Because the instruction and assessment tools are changed very often, there is a need to outline in detail the activities and criteria promoting the global, contextual, professional, technical, and soft competencies. Several endeavors were suggested by engineering education researchers for how to build up reliable and valid assessment tools for professional competencies. These can be said a set of common competencies that all engineering students should have by the time they enter the work force or few months after they got in the business organizations. Yet, fewer such efforts have been put forth to recommend practices for promoting and developing students' professional competencies, especially in the skill-based learning context. In an effort to address this gap, Evans et al. (2015) outlined an initiative for incorporating technical and soft (professional) skills as a part of the curriculum in a competency-based setting whose primary goal is to improve communication skills and social and cultural awareness. Although the current literature suggests that CBSL is an effective educational method, more empirical research is needed to validate the outcome of findings. There is a lack of empirical investigation in the current literature quantifying the effects of CBSL and other pedagogical approaches, instructional methods, and assessment tools on student success (learning level) at course-base and program-base in a reliable and valid manner. In order to have confidence in causal claims about the effectiveness of CBSL, its definition and implementation need to be consistent across programs. Currently, the definition and extent to which CBSL is implemented across programs vary widely. This lack of consistency and control may reduce the confidence in the causal claims by which CBSL leads to an improvement in SO. For the assessment of learning, it is important for all future researches to consider past findings. Integration of previous research is necessary for advancement of the science of learning.

Because of the nature of student learning, it is usually difficult or even sometimes impossible to get the adequate or complete information for SO. Incomplete information is the fundamental reason for a learning problem of being "grey" which is decided based on the amount of available information. The grey elements, the grey relations represent the relations with incomplete information that an evaluator might have. The grey relational analysis (GRA) uses data based on the level of similarity and variability among all factors to establish their relation and compare them quantitatively. This is unlike the traditional statistical methods handling the relation between student learning-level assessment tools which require plenty of statistical data. GRA requires less data and can analyze many factors that can overcome the disadvantages of statistical method. Grey theory is similar to fuzzy set theory. It is an effective mathematical tool to deal with systems analysis characterized by imprecise and incomplete information. The theory is based on the degree of information known. The advantage of grey theory over fuzzy theory is that grey theory takes into account the condition of fuzziness; that is, grey theory can deal flexibly with the fuzziness situations.

The assessment approaches that are based on fuzzy linguistic terms can add more understandable meanings to the numerical outcomes of SO. The student's learning performance will be considerably more understandable by the educators, instructors, and the parents. We can define several fuzzy linguistic terms for defining student learning performance. Some of them are as follows: *poor, average, satisfactory, good, excellent, and outstanding.* Normally, these expressions are used in our daily life to define the quality of products or services. In contrast, in statistical methods, different scores of each assessment tool are added up based on determined weights to obtain a single score for an individual student's performance. Thus, marks assigned by an evaluator are only approximations. Conversely, the academic performance assessment involves the measurement of ability, competence, and skills which are the concepts that have vague meanings and can be approximated by fuzzy linguistic terms. Eventually, linguistic terms can be awarded to a single student's achievement as well as a group of students who had already taken a course. In conclusion, the performance assessment yields feedback on the effectiveness of teaching and the extent of achievement of course goals, and offers information on the effectiveness of the learning in the end.

2.6.1 *AI methods for prediction and assessment of student's learning at course level*

Fuzzy systems and artificial neural networks (ANNs) are soft-computing approaches for modeling human behaviors. Neural network-based systems can be trained by numerical data and fuzzy rules can be extracted from neural networks. Similarly, fuzzy rule-based classification systems can be designed by the linguistic knowledge. Fuzzy systems are appropriate if sufficient expert knowledge about the problem is available while neural systems are useful if sufficient course data are available. Both approaches build nonlinear systems based on bounded continuous parameters, but the difference is that neural systems are treated in a numeric quantity, whereas fuzzy systems are treated in a symbolic qualitative manner. The integration of neural and fuzzy systems leads to a symbolic relationship in which fuzzy systems provide a powerful framework for expert knowledge representation while neural networks have learning capabilities. The aim of integration is to build more intelligent decision-making systems. Fuzzy systems are presented in the form of fuzzy rule base and the most important area in the application of fuzzy set theory. Designing a fuzzy rule-based system involves derivation of desired "IF-Then" fuzzy rules, partitioning of universes, and addressing of the membership functions (MFs). A few systematic design procedures are available for the designing intelligent models for prediction though fuzzy systems have reached a recognized success in several significant application areas such as control problems, modeling, performance classification, and decision systems. In recent years, it has become clear that neural and fuzzy hybrid systems have advantage in the application of performance assessment and several other areas. ANNs are also used for prediction of certain assessment problems.

Combination of fuzzy sets with neural networks is called adaptive neuro-fuzzy systems (ANFIS) that aims developing an intelligent model with learning capability. Both neural networks and fuzzy systems are dynamic and parallel processing systems that use input–output data to produce an output. The main features of an ANFIS model are as follows:

- to process data according to fuzzy reasoning mechanisms,
- to recover encoded knowledge in the form of fuzzy rules and use them to establish the model,
- to learn rules covering the whole input space and have universal approximation characteristic,
- to process the data and automatically determine the ranges of MFs.

The neural networks learn in two main phases. In the first phase, the network performs a self-organization to determine the rules and the respective MFs from the data set. In the second phase, they learn consequent parameters using a supervised scheme. On the other hand, the linguistic variables and their labels are the backbone of fuzzy rule-base systems and the fuzzy models. Schmucker (1984) defined a linguistic variable as a variable whose values are linguistic words or sentences in a natural or synthetic language. For instance, *He has learned very well and did excellently in the final exam* is a vague and imprecise linguistic statement in the real life since someone may ask; *how very well he/she has learnt?* To answer these kinds of questions, a systematic approach is needed. There is an increasing interest to augment fuzzy systems with learning and adaptation capabilities. This enhancement effort is succeeded by hybridizing the approximate reasoning method of fuzzy systems with learning capabilities of neural networks. The merits of neural networks and fuzzy systems can be integrated in a neuro-fuzzy approach. Eventually, the combination of neural networks and fuzzy systems (neuro-fuzzy systems) has been recognized as a powerful alternative approach to develop adaptive soft-computing systems.

The student learning at a course level can be determined by an ANFIS model that is a multi-input single-output (MISO) system. This model will have certain assessment parameters and one output attribute. The input attributes of the ANFIS model are the parameters such as *Quizzes, Majors, Midterm exam, Final exam, Active learning sessions, and Term project* and the output is the *Student's learning level.* These imprecise attributes are called as fuzzy linguistic variables and used commonly in an educational system. They are expressed by premise parameters (fuzzy linguistic term) in daily life such as *very poor, poor, unsatisfactory, average, good, very good, and excellent.* Arithmetical and statistical methods are unable to offer an effective inference procedure to assess the learning level of a student in a more natural way using linguistic variables or terms. The achievement (learning level) of a student is usually evaluated and presented by numerical values in the current educational applications; however, linguistic terms are more natural than them. Law (1995) expressed the natural linguistic terms *(e.g., unsatisfactory, average, good, very good, excellent, etc.)* to represent the student academic performance by ignoring mostly their inherent nature of vagueness. However, the academic performance evaluation involves the measurement of ability, competency, and skills that are uncertain and imprecise concepts in fuzzy systems. The use of a natural language such as *good, very good, and excellent* would yield more meaningful explanations than the current educational applications and could allow flexibility in reasoning and judgment of the student's learning level.

2.6.2 Existing methods for learning performance assessment in education

Education is a process of accepting students from a low level of standing, drives them through several stages of development, and produces individuals qualified with certain abilities, skills, and attributes who are fit for a job or a higher level in education. Engineering courses are indispensable elements of the educational learning process. The main steps in the education process are specified as the curriculum. Traditionally, the features of overall system are defined by models that are specified by a set of courses and students have completed and convince the educators that they have learnt the subjects specified by a set of content of courses accumulated in them. This approach, however, does not guarantee that students will achieve the qualifications required and the goal of learning level. Therefore, educators have stipulated a new approach that relies on what the students have learnt rather than what he/she has been taught. An act of assigning a qualitative or quantitative merit to student achievement is defined as the academic assessment. Assessment of a student's academic performance is the assessment of their learning level and it is one of the most central practice used for three main reasons:

- to decide on pass and failure of a student from a course,
- to obtain the indication of a student's learning level, and
- to provide information on the effectiveness of teaching.

Engineering graduates will implement the competency, technical knowledge, and soft skills gained during their employment in work places. In the current engineering practices, the half-life of learning process is around 5 years, which means that students can only retain half of what they have learnt at the university in 5 years of education period. Hence, they have to re-new themselves and must not cut their relation with information flow toward them. The pace of knowledge renovation is very high in the current era. Therefore, the educational objectives of the engineering programs must be clearly stated. Soft and technical skills, attributes, and the other abilities must be gained by students at the graduation to fulfill the student learning objectives (SO); hence, they should be stated and related with courses learning objectives. In this respect, student learning level is the most likely measure to verify achievement of course-learning objectives to expose the effectiveness of learning environment and to monitor standards of the educational organization. In statistical methods, a student's learning level is evaluated based on the marks collected in a course during a semester. This approach has numerous restrictions in satisfying the students learning-level measurements for an outcome-based assessment approach. It can be classified into numerous categories such as single numerical scores usually referring to 100%, single letter grades (e.g., A, B, C, D, or F), nominal scores (e.g., 1, 2, 3, ...10), linguistic terms such as "Fail", or "Pass" or single grade-points from 0.00 to 4.00 (Taylan et al., 2017). In the literature, assessment techniques have been widely discussed. For instance, Ratcliff et al. (1997) worked on qualitative and quantitative assessment techniques. Similarly, Lopes et al. (1997) studied the qualitative techniques based on several indicators for teaching, research, and quality of a department. They also declared that the indicators of "quantities" have fuzzy meanings. Deniz and Ersan (2002) presented several ways by which student's learning level can be analyzed and presented as an academic decision-making process and a decision-support system can be used for its assessment. Ma and Zhou (2000) presented an integrated fuzzy set approach to assess the outcomes of a student learning. They exploited fuzzy set principles to represent the imprecise concepts for subjective judgment and applied a fuzzy set method to determine the assessment criteria and their corresponding weights. Their study aimed at encouraging students to participate in the whole learning process as well as providing an open and fair assessment environment. Ebel and Prade (1991) employed formative and summative assessment forms to evaluate the academic performance. Teachers, educational planners, and even students need feedback of learning level during the course of instruction. Hence, the formative assessment is conducted to monitor the progress of instruction through series of tests, quizzes, and observations. The summative assessment is carried out at the end of each instructional segment through tests and final examinations to provide information on how much students would achieve their objectives. However, the assessment is a part of learning process and not something to be done after the instruction is completed (Taylan et al., 2017). Also, written examinations at the end of the semester may not provide a complete picture of what the students have learnt. Preferably, a combination of both assessment methods may be used to provide a full coverage of important learning outcomes. Eventually, in an educational system, assessment tools may consist of series of tests, quizzes, portfolios, formal written examinations, individual assignments and course works, group works, observations, projects, publishable materials, and oral presentations (Taylan et al., 2017). The commonly used technique for assessment of a student learning performance relies on statistical methods that award numerical values or linguistic labels to a given piece of student's work. These values and labels have been frequently used to represent the student's achievement without exploiting any other alternative to double check the student's performance. The grade awarded by an evaluator could be an approximation only because he/she typically assigns a numerical score to the student's effort and usually does not consider the competency, a comparative evaluation, skills obtained. On the other hand, judgmental assessment and linguistic reasoning are not used to reward the student's learning effort (Taylan et al., 2017). The numerical scores may vary according to criteria of different evaluators because their sensitivities, experiences, and standards may be different.

2.6.3 *Adaptive neuro-fuzzy inference system (ANFIS) for student learning level assessment*

An ANFIS is a computing framework based on the concepts of fuzzy sets, fuzzy If-Then rules, and fuzzy reasoning that have been successfully used in a wide variety of applications. The ANFIS can utilize linguistic information from human experts, simulate human thinking procedures, and approximate an inexact nature of the real world. The outcomes of a fuzzy model are usually fuzzy values defined by a degree between 0 and 1. A fuzzy model can produce crisp numerical outcomes.

Developing an ANFIS model to evaluate the student learning level requires to take into account scarcity of data and style of input space partitions. The common approach is to divide the data set into training, testing, and checking sets. Training data set serves in model building while the testing data set avails for the validation of the model. The ANFIS model is able to produce crisp numerical values and includes the following parameters: input and output variables by linguistic statements, fuzzy partition of the input and output spaces, choosing the MFs for the input and output linguistic variables, deciding on the types fuzzy control rules, designing the inference mechanism, and choosing a de-fuzzification procedure. In a fuzzy model, the number of MFs and rules has a critical effect on how fine a control level can be achieved in the fuzzy model. The number and type of MFs are decided through application of subtractive clustering algorithm in which each data point belongs to a cluster at a degree specified by a membership grade. The idea of this clustering algorithm is to divide the output data into fuzzy partitions that overlap with input parameters. Thus, an ANFIS model is developed based on the past numerical data

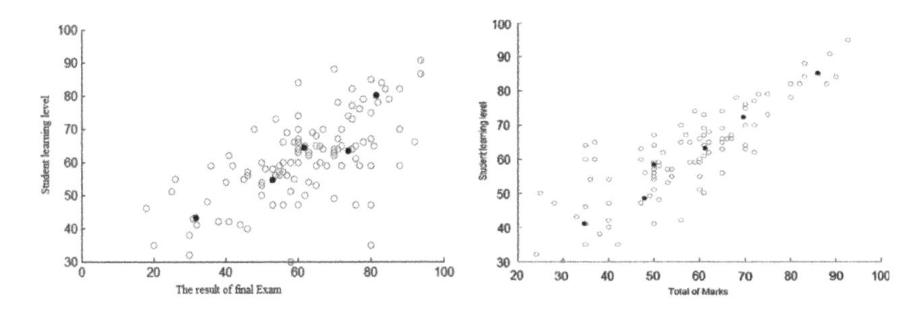

Figure 2.1 Clustering for the constitution of rule-based system.

of the student performance achieved during previous semesters. The least-squares method (LSE) can be employed in the inner loop of ANFIS to fine-tune parameters and to optimize the number of rules.

Fuzzy clustering approach is used to produce optimal number of rules that are based on the clustering of input and output data sets. In this method, the number of rules is equal to the number of clusters regardless of the number of input variables. It assumes each data point as a potential cluster center and calculates a measure of likelihood defining the cluster center based on the density of surrounding data points. To generate an ANFIS model, a cluster radius must be specified to indicate the range of influence of the cluster. Figure 2.1 illustrates the structure of rule-based system used for a data space. Specifying a small cluster radius yields many small clusters in the data, resulting in a lot of rules.

2.6.4 Fuzzy reasoning for determining student learning level

Fuzzy reasoning is the backbone of fuzzy inference system (FIS) and it is one of the most important modeling tools based on fuzzy sets and systems. Fuzzy reasoning is an inference procedure to derive conclusions from a set of fuzzy if-then rules and facts. Figure 2.2 shows the reasoning procedure for a first-order Sugeno and Kang (1988) fuzzy model. Since each rule has a crisp output, the overall output can be obtained via a weighted average; in this way, a time-consuming process can be avoided.

Schmucker (1984) stated that a linguistic variable is a word or sentence in a natural or synthetic language. Such statements as *He has learned very well and did excellently in the final exam* are really vague linguistic terms in real life. The ANFIS under consideration is a MISO model and has six inputs and one output parameter. These parameters are called fuzzy linguistic variables that are very common in an educational system. However, they are imprecise, vague, and incomplete fuzzy terms. They might be introduced and expressed by premise parameters (fuzzy linguistic terms) such as *poor, average, good, excellent, and outstanding,* as given in Figure 2.3.

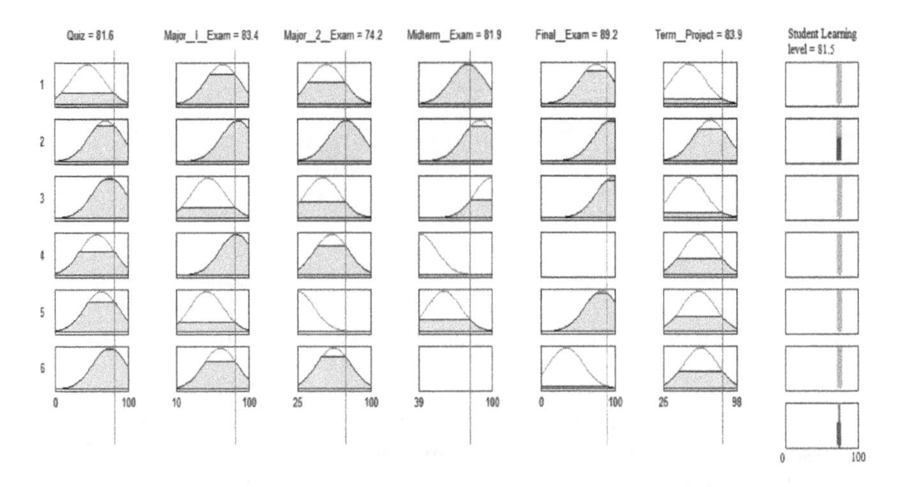

Figure 2.2 Fuzzy reasoning procedure for assessment of student learning at course level.

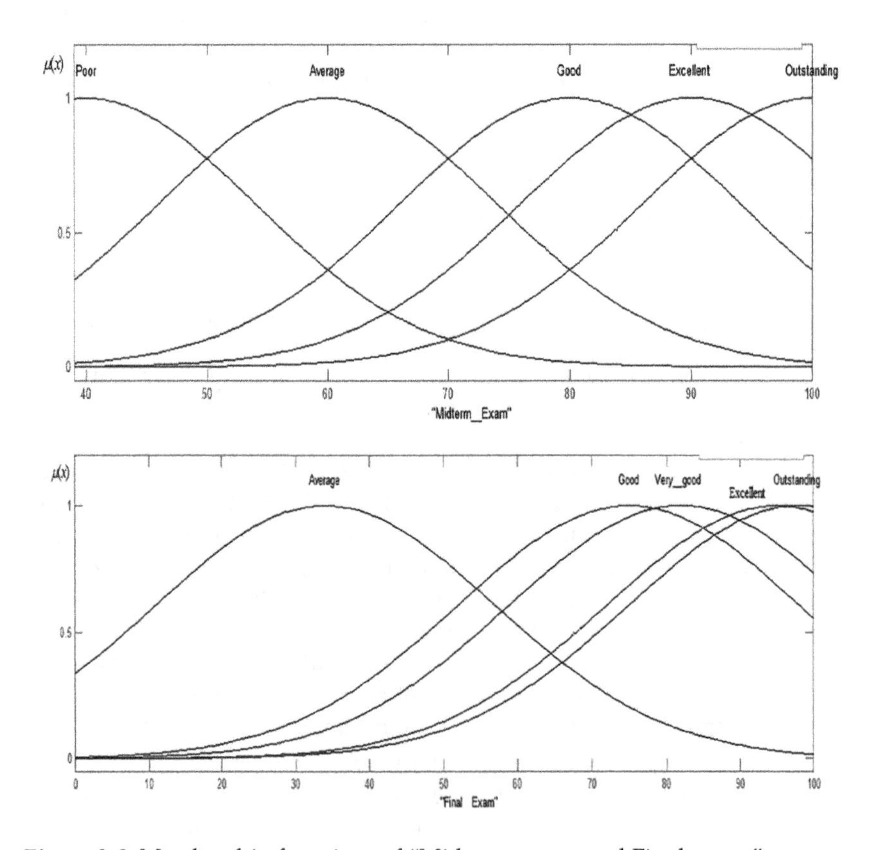

Figure 2.3 Membership functions of "Midterm exam, and Final exam."

Figure 2.3 illustrates fine-tuned MFs of input variables; *final exam* and *midterm exam* versus the output variable *student learning level*. The combination of fuzzy sets will characterize the membership degree of a student achievement. Fuzzy rules are mathematical relationships mapping the relations between inputs and output parameters constituted from fuzzy linguistic variables and their term sets. Conventional techniques for system analysis are not suited for dealing with a humanistic system whose behavior is strongly influenced by human judgment, perception, and emotions. This belief yields the concept of linguistic variables as an alternative approach to model human thinking. Fuzzy *If-Then* rules are known as fuzzy implications or fuzzy conditional statements that are widespread in our daily linguistic expressions. As some fuzzy rules are presented below, they are the backbone of a fuzzy model.

For example, the following are the complete fuzzy rules defining the relations of assessment tools and student learning level:

IF the Quiz of a student is Average and Major exam is Good and Midterm exam is Very-good and Final exam is Good THEN the Student learning level is Good

IF the Quiz of a student is Poor and Major exam is Average and Midterm exam is good and Final exam is Poor THEN the Student learning level is Poor

Fuzzy *If-Then* rules are constituted from two parts: antecedent parts and consequence part. A typical fuzzy rule in a fuzzy model has the form presented above where fuzzy terms are fuzzy linguistic values defining the linguistic variables. *Quiz exam is poor* and *Final exam is good* are antecedent part of a rule. Similarly, *Student learning level is average* is a conclusion or consequent part of a rule set.

An adaptive network is a self-learning algorithm consisting of several nodes connected through directional links. Each node transfers a function on its incoming signals to generate a single output. Each link specifies the direction of signal flow from one node to another. All or a part of nodes are adaptive, which means that the outputs of these nodes depend on modifiable parameters pertaining to the nodes. The learning rule specifies how these parameters should be updated to minimize a prescribed error measure, which is a mathematical expression determining the discrepancy between the network's actual output and desired output. The parameters associated with the MFs will change through the learning process. The computation of MFs' parameters is facilitated by a gradient vector, which provides a measure of how well the ANFIS would be modeled using the input/output data for a given set of parameters. Once they are constituted, any of several optimization routines can be applied in order to adjust the parameters to reduce the error.

The ANFIS captures the essential components of underlying dynamics and the training data contain effects of the initial conditions which might not be easily accounted for by the essential components identified by ANFIS. The crisp results of fuzzy model that are representative for student learning level and the quality distribution of the student learning level are presented in Figure 2.4. The quality assessment of student learning level is carried out by linguistic terms such as *very poor, poor, average, good, very good, excellent, and outstanding* which identify the student learning level substantially.

As seen in Figure 2.4, the scatter plot of 110 principal component outcomes of the ANFIS model is depicted. Normally, each outcome represents the learning-level achievement of a student in a quality category and has a crisp numerical value representing the quality of learning level for a group of students in a course. This assessment method will allow more flexible and judgmental assessment on a group of student performance in a course, and it is more understandable by educators, instructors, and parents. The student's learning level and their performance achievement are expressed by linguistic quality indicators. In our case, we have used seven linguistic terms such as *very poor, poor, average, good, very good, excellent, and outstanding* to identify the quality distribution of learning level. Normally, these expressions are used in our daily life to grade the quality of products or services. For instance, the students who achieved 90 and above score are considered as a successful group, and their performance can be expressed by a fuzzy linguistic value *outstanding*. This linguistic value is a score showing that a student in this group achieved an outstanding learning level. Similarly, students achieved a score of learning level between 50–60 interval, are in the same group, and their learning level can be expressed poor linguistically. Performance data of previous and current students in a computer system might be an excellent source to achieve

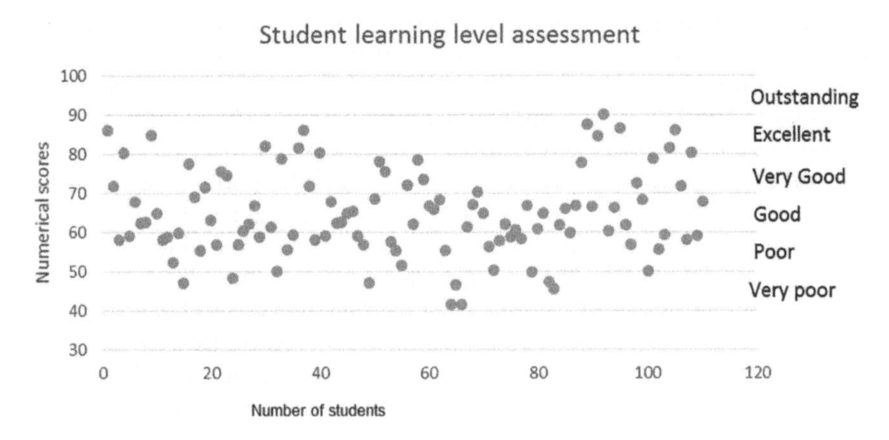

Figure 2.4 Outcomes of ANFIS model and their linguistic values at course level.

satisfactory results in evaluation of the students' learning skills, and levels to bring out the quality of educational system. In this approach, a student who received 84 in his total score out of 100 will partially be member of *excellent* set with a membership degree of 0.85 and will also be member of *outstanding* set with a membership degree of 0.56. Similarly, a student achieving 56 out of 100 in his/her total score will be member of *poor* set with a membership degree of 0.9, and will also be member of *average* student set in his/her learning level with a membership degree of 0.5. The same student will in the set of *very poor* with a membership degree of 0.35.

In conclusion, there are several reasons to assess students' learning performance. This assessment gives feedback on the effectiveness of teaching approach, the extent to which the course aims could be achieved, competency-based learning, and knowledge on the effectiveness of the teaching–learning process. Although different teaching–learning assessment methods have been employed in primary, secondary, and tertiary education, it is believed that there is still a lack of empirical study in the current literature quantifying the effects of CBSL, instructional methods, and assessment tools on SO and program success in a reliable and valid manner. Currently, the definition and extent to which CBSL is implemented across programs vary widely. This lack of consistency and control may reduce the confidence in the causal claims that student learning leads to an improvement in SO.

For the assessment of learning at engineering program level, it is important for all future researches to consider past findings and use them for the assessment of future global engineering education assessment. In different countries, statistical methods are mainly employed for the learning assessment. However, in existing evaluation methods, evaluators usually lack a formal reasoning mechanism to support the inference. Typically, numerical outcomes (marks) are used for decision-making according to given marking schemes, experiences, sensitivities, and standards. The grade point average (GPA) which is used for the assessment of a student and the marks appearing in GPA that are assigned by an evaluator are only the approximations. However, the academic performance and the learning-level evaluation involves the measurement of abilities, competency, and soft and technical skills that can be approximated by fuzzy logic approaches. Eventually, linguistic terms can be awarded to a single student's achievement as well as a group of students who had already taken a course.

2.7 GRA for optimization of student learning assessment

In this section, a GRA supported fuzzy subtractive clustering approach was used to generate objective number of rules which was based on the clustering of student learning-level assessment tools. In order to generate the inference structure, a cluster radius was specified to indicate the range of influence of the cluster. This process is supported by GRA to reduce the error between each cluster center and each data point belongs to the cluster. These cluster centers are considered as the centers of the fuzzy rules' premise in a zero-order Sugeno fuzzy model. As a result, five cluster centers were determined for the given data set of SO (see Figure 2.5). The inference system under consideration is a MISO system which has six input parameters (I_{1-6}) and one output (O) parameter.

The output is the *Student learning quality level (O)*. The assessment tools are called the written exams and the other activities in an educational system. The GRA used in this section, is described through the following steps:

- availability of key imprecise attributes.
- the number of attributes is known and limited.
- each attribute is measured independently.

The aggregated distances between the categorical data are then processed together using GRA to reduce uncertainty. A significant modification was

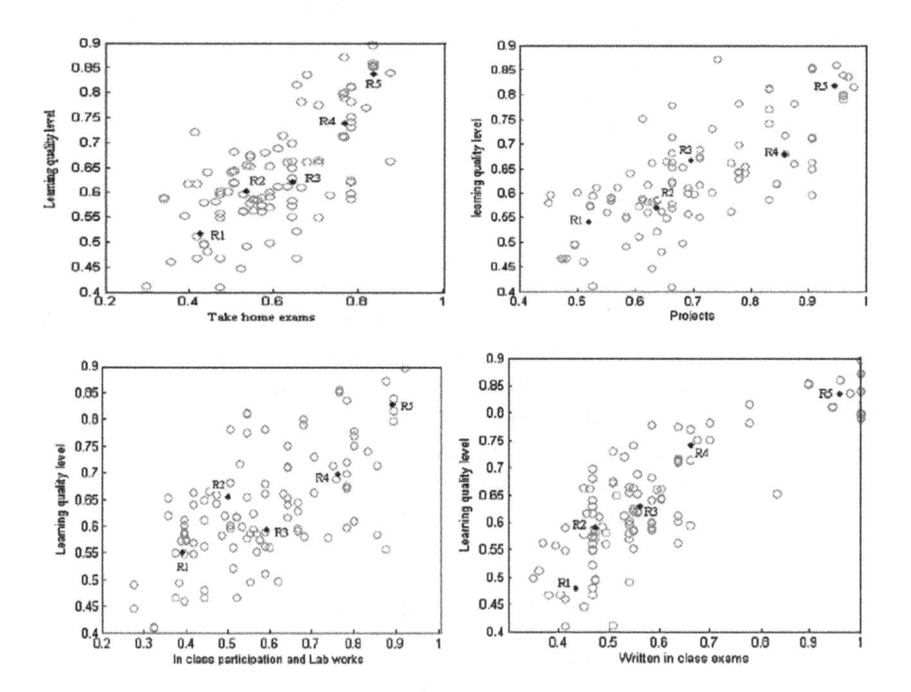

Figure 2.5 Distribution of student learning performance for different assessment tools.

carried out to incorporate the fuzzy distance between two numeric values. The aim of this new distance is to reduce uncertainty associated with similarity measurement at the *k*th points. The approach proposed has several limitations: first, the similarity degree between two different values falling in the same fuzzy set is equal to 1. The proposed fuzzy distance in this study attempts to overcome the challenges by proposing more general formula based on the concept of fuzzy set theory that can be applied to different MFs. Table 2.2 shows the students' mark and their learning level obtained from statistical assessment. Table 2.3 depicts the grey relational coefficients and the grey outcomes of student learning level.

Table 2.4 depicts the center of learning levels, determined for fuzzy rules and MFs. Each center shows a group of students and their performance in terms of their learning level which can be presented by MFs.

2.7.1 Optimization of student learning model

In this section, we employed the Back-propagation neural networks (BPNNs) algorithm for learning from the given data set, and reduce the errors by propagating them to optimize the FIS. This neural network algorithm is a self-learning algorithm consisting of several nodes connected through directional links.

The student learning assessment tools are arrayed into a vector expressed by $f_j(I_1, I_2, I_3, I_4, I_5, I_6, O)$. Given a training set of input–output pairs, the algorithm provides a procedure for changing the weights in a back-propagation network to classify the given input patterns correctly. The basis for this weight updating algorithm is simply the gradient–descent

***Table* 2.2** Students' mark depicting the learning level

I_1	I_2	I_3	I_4	I_5	I_6	Learning (*O*)level
16	12	20	15	5	14	15
25	16	20	45	30	30	29
59	32	60	0	10	50	43
22	16	17	45	35	10	20
60	34	43	25	30	39	41
15	35	35	45	0	44	33
70	40	24	15	5	38	37
42	15	40	60	20	39	38
22	10	17	25	15	10	16
45	40	29	30	30	33	35
36	52	75	40	20	40	41
25	33	54	50	25	50	41
72	24	43	40	25	50	48

Table 2.3 Grey relational coefficients and grey relational marks

A1	A2	A3	A4	A5	A6	Learning level
0.9836	0.8372	0.9074	0.6667	0.7778	0.8974	0.853
0.8571	0.7659	0.9074	0.4000	0.3684	0.6364	0.713
0.5769	0.5714	0.5213	1.0000	0.6364	0.4667	0.575
0.8955	0.7659	0.9608	0.4000	0.3333	1.0000	0.797
0.5714	0.5538	0.6364	0.5455	0.3684	0.5469	0.587
1.0000	0.5455	0.7101	0.4000	1.0000	0.5072	0.674
0.5217	0.5070	0.8448	0.6667	0.7778	0.5556	0.618
0.5607	0.7826	0.6622	0.3333	0.4667	0.5469	0.623
0.8955	0.8780	0.9608	0.5455	0.5385	1.0000	0.84
0.6667	0.5070	0.7778	0.5000	0.3684	0.6034	0.644
0.7404	0.4337	0.4495	0.4286	0.4667	0.5385	0.579
0.8571	0.5625	0.5568	0.3750	0.4118	0.4667	0.585
0.5128	0.6545	0.6364	0.4286	0.4118	0.4667	0.521

Table 2.4 The cluster centers and geometric location of fuzzy rules

	R1	R2	R3	R4	R5
(I_1, O_1)	(0.39,0.55)	(0.52,0.66)	(0.60,0.58)	(0.78,0.69)	(0.91,0.84)
(I_2, O_2)	(0.42,0.52)	(0.54,0.60)	(0.65,0.62)	(0.78,0.74)	(0.85,0.84)
(I_3, O_3)	(0.52,0.54)	(0.62,0.57)	(0.71,0.66)	(0.87,0.68)	(0.96,0.82)
(I_4, O_4)	(0.38,0.41)	(0.43,0.51)	(0.61,0.68)	(0.71,0.76)	(0.91,0.82)
(I_5, O_5)	(0.57,0.61)	(0.64,0.68)	(0.73,0.79)	(0.77,0.84)	(0.92,0.83)
(I_6, O_6)	(0.44,0.47)	(0.48,0.59)	(0.58,0.63)	(0.67,0.74)	(0.96,0.84)

method and used for perceptrons with differentiable units. For a given input–output pair *(Is, O)*, the back-propagation algorithm performs two-phase data flow. In order to eliminate the problem, GRA supported fuzzy model was employed. This data set verifies the capability of the resulting FIS model and provides an unbiased index for selecting the robust model and its parameters.

Fuzzy linguistic values such as *Poor(P), Average(A), Satisfactory(S), Good(G), Excellent(E), and Outstanding(O)* were defined for assessment tools. *(I)s* are the fuzzy linguistic variables (assessment tools) such as "Final exam, Major Exams, Quizzes and etc," The developed procedure is able to produce crisp outcomes for any student learning level between 0 and 100. Figure 2.7 indicates the Sugeno reasoning procedure for our fuzzy input variables. Some linear MFs depicting the learning level of students are presented below. For instance, Figure 2.6 shows linear MFs of student learning assessment tools such as "in class participation & lab works, and "take home exams."

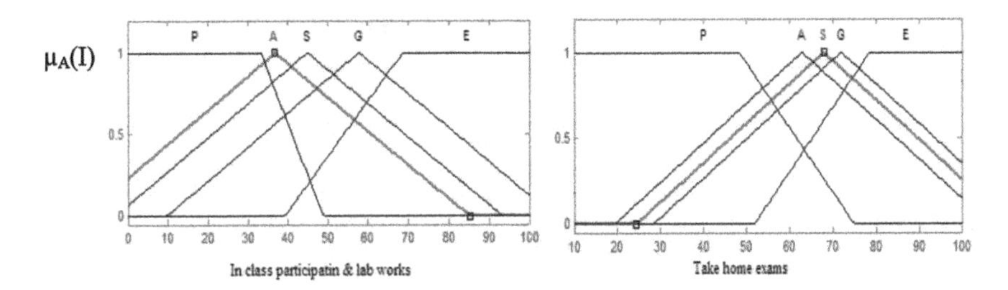

Figure 2.6 Fine-tuned linear MFs for "in class participation & lab works," and "take home exams."

$$\text{Triangular}(I_1;10,60,100)=\begin{cases} 0, & I_1 \le 10 \\ \dfrac{I_1-10}{50}, & 10 \le I_1 \le 60 \\ \dfrac{100-I_1}{40}, & 60 \le I_1 \le 100 \\ 0, & 100 \le I_1 \end{cases}$$

Figure 2.6 illustrates graphical representations of fine-tuned MFs $(\mu(I))$ of input variables *in class participation & lab works* (I_1) and *take-home exams* (I_2) versus the output variable (*O*), and the linguistic term set used in this study. The combination of fuzzy sets will characterize the membership degree $\mu_A(I)$, where *I* is a score (mark) representing the achievement of a student learning level, e.g., I_1*(in class participation & lab works)* = 60. A fuzzy set is characterized by its MF which is conveniently expressed by a mathematical formula. The mathematic of the triangular MF for linguistic term *Good (G)* might be constituted as it is given in equation above for this study. A triangular MF is specified by three parameters such as {10, 60, 100} for linguistic term *Good*. The MF of other fuzzy linguistic values can be constituted in a similar manner.

Model outcomes are validated by randomly selecting and presenting crisp input values to the trained FIS model given in Figure 2.7, in order to see how well the FIS model predicts the corresponding data set of the output values. The developed FIS model is expected to create outcomes associated with the minimum testing errors. As can be seen in Figure 2.8, data were used for identification of each assessment tool and 110 data were used for testing the model. This data set contains all the necessary representative features of the assessment tools. The outcomes obtained are very encouraging for assessment of learning quality level. A part of crisp outcomes of fuzzy model representing the quality of learning level is given in Table 2.5.

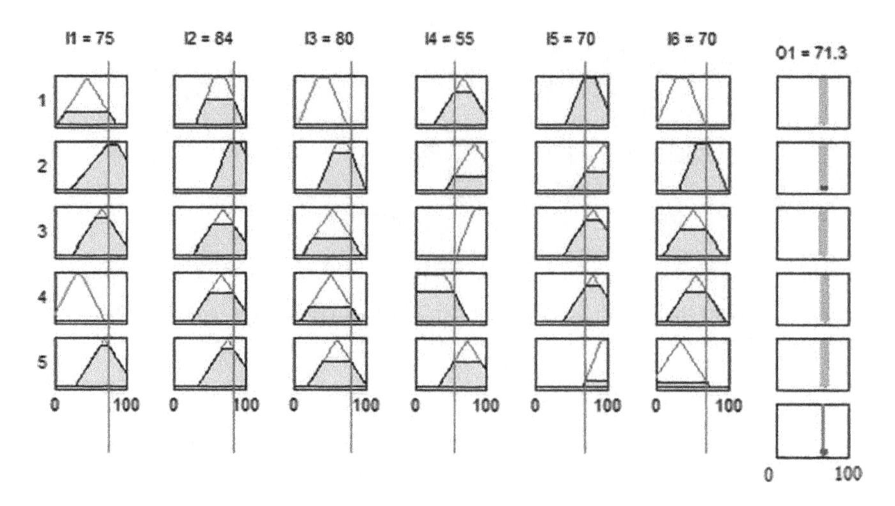

Figure 2.7 Fuzzy reasoning procedure for Sugeno model of SLP assessment.

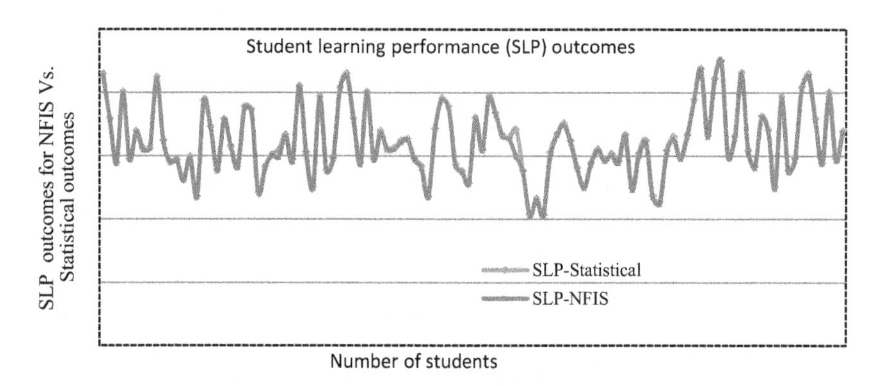

Figure 2.8 The Leaning level outcomes of statistical method vs. FIS for different students.

2.8 Results and discussion

This study presents an ANFIS approach to evaluate student's learning performance using the outcomes of various assessment processes. The assessment outcomes are obtained from the ANFIS approach and compared with those of the statistical method outcomes. For example, if a student gets $I_1 = 75$, $I_2 = 84$, $I_3 = 80$, $I_4 = 55$, $I_5 = 70$, and $I_6 = 70$ in one of his courses, the outcome of this student is $(O_2) = 71.9$ according to statistical method. However, different evaluators might achieve different results according to given weights for such a student. As can be seen in Figure 2.7, the ANFIS yields $O_2 = 71.3$ outcome for the same assessment tool's value. Similarly, a set of outcomes of ANFIS are given in Figure 2.8. It is clear

Table 2.5 Outcomes of statistical method and FIS approach for some randomly selected

Q	*M*	MD	*P*	*S*	*F*	Statistical SAP outcomes	ANFIS SAP outcomes, subclustering	ANFIS SAP outcomes, grid partitioning
60	48	25	60	80	60	50.85	50.8502	51
50	67	46	50	75	50	55.3	55.3001	56
60	76	57	60	75	50	55.2	55.2001	55.9999
85	70	46	85	75	53	56.7	56.6999	57
60	45	32	60	75	50	44.5	44.5	41
90	87	75	90	95	72	76.85	76.8496	77.0001
95	72	65	95	80	62	67.2	67.2	68
67	38	50	67	75	61	53.7	53.7	49.9999
85	73	60	85	80	72	69.5	69.5	69.9999
85	75	71	85	90	50	63.25	63.25	65.4999
67	66	70	67	65	70	63	63	63

that the outcomes of ANFIS are as robust as the statistical method, if the ANFIS model is trained properly. The developed ANFIS has been tested for 110 principal input values (marks obtained by different students) and the outcomes of them (Student learning level) are presented in Figure 2.7. Similarly, Figure 2.8 presents a comparison between the outcomes of statistical method and ANFIS approach. The outcomes of both methods are very close to each other. This assessment is to categorize the student using fuzzy linguistic terms and develop clusters for their learning performance. The clusters might be developed for last 5 or even for 10 years period, and the teaching performance of an instructor, the learning performance of students, or the quality of lecturing might be evaluated for further actions.

The students achieved scores of 80 and above might be considered as a cluster and the learning performance of them might be expressed by the linguistic term *Excellent (E)*. Similarly, the students achieved scores between 70 and 80 might be considered as a cluster and the learning performance of them might be expressed by the linguistic term *Good (G)*. These linguistic values are the fuzzy terms presenting the student's learning achievement and learning quality level. Similarly, students scored between 50 and 60 may represent a group with a reasonable level of achievement and their performances might be expressed linguistically as *Average (A)* and so on. This flexibility does not exist in the statistical approaches. The assessment approaches that are based on fuzzy linguistic terms add more considerable and understandable meanings to the numerical outcomes of models. Hence, the student's learning performance is now more understandable by the educators, instructors, and the parents.

We have defined five fuzzy linguistic terms in this study: *Poor (P), Average (A), Satisfactory (S), Good (G),* and *Excellent (E).* Normally, these expressions are used in our daily life to define the quality of products or services. In contrast, in typically used statistical methods, different scores of each assessment tool are added up based on determined weights to obtain a single score for an individual student's performance. Thus, marks assigned by an evaluator are only approximations. Conversely, the academic performance assessment involves the measurement of ability, competence, and skills which are fuzzy concepts and can be approximated by fuzzy linguistic terms. Eventually, linguistic terms can be awarded to a single student's achievement as well as a group of students who had already taken a course. In conclusion, the performance assessment yields feedback on the effectiveness of teaching and the extent of achievement of course goals, and offers information on the effectiveness of the learning in the end.

References

ABET (2016). Retrieved from http://www.abet.org/accreditation/accreditation-criteria/accreditation-policy-and-procedure-manual-appm-2016–2017.

Aeschlimann, B., Herzog, W., & Makarova, E. (2016). How to foster students' motivation in mathematics and science classes and promote students' STEM career choice. A study in Swiss high schools. *International Journal of Educational Research,* 79, 31–41.

Al-Thani, S. J., Abdelmoneim, A., Daoud, K., Cherif, A., & Moukarzel, D. (2014). A perspective on student learning outcome assessment at Qatar University. *Quality in Higher Education,* 20, 255–271.

Badilla Quintana, M. G., Carrasco Saez, J. L., & Prats Fernandez, M. A. (2014). Use of PLE-portfolio to assess the competency-based learning through Web 2.0 in technical engineering education. *International Journal of Engineering Education,* 30(3), 675–682.

Bloom, B. S. (1984). The 2 sigma problem: The search for methods of group instruction as effective as one-to-one tutoring. *Educational Researcher,* 13(6), 4–16.

Bogue, B., Shanahan, B., Marra, R. M., & Cady, E. T. (2013). Outcomes-based assessment: Driving outreach program effectiveness. *Leadership and Management in Engineering,* 13(1), 27–34.

Canaleta, X., Vernet, D., Vicent, L., & Montero, J. A. (2014). Master in teacher training: A real implementation of active learning. *Computers in Human Behavior,* 31, 651–658.

De Los Rios-Carmenado, I., Rodriguez Lopez, F., & Perez Garcia, C. (2015). Promoting professional project management skills in engineering higher education: Project-based learning (PBL) strategy. *International Journal of Engineering Education,* 31(1), 184–198.

Deniz, D. Z., & Ersan, I. (2002). An academic decision-support system based on academic performance evaluation for student and program assessment. *International Journal of Engineering Education,* 18(2), 236–244.

Díaz Lantada, A., Lafont Morgado, P., Muñoz Sanz, J. L., Muñoz Guijosa, J. M., Echávarri Otero, J., Chacón Tanarro E., & De la Guerra Ochoa, E. (2013a). Study of collaboration activities between academia and industry for improving the teaching learning process. *International Journal of Engineering Education*, 29(5), 1059–1967.

Díaz Lantada, A., Lafont Morgado, P., Muñoz Sanz, J. L., Muñoz Guijosa, J. M., Echávarri Otero, J., Chacón Tanarro E., & De la Guerra Ochoa, E. (2013b). Towards successful project based teaching-learning experiences in engineering education. *International Journal of Engineering Education*, 29(2), 1–15.

Díaz Lantada, A., Munoz-Guijosa, J. M., Tanarro, E.C., Otero, J. E., & Sanz, J. M. (2016). Engineering education for all: Strategies and challenges. *International Journal of Engineering Education*, 32(5(B)), 2255–2271.

Di Trapani, G., & Clarke, F. (2012). Bio techniques laboratory: An enabling course in the biological sciences. *Biochemistry and Molecular Biology Education*, 40(1), 29–36.

Ebel, R.L., & Prade, H. (1991). *Essentials of Educational Measurements*. 5th ed., Englewood Cliffs, NJ: Prentice Hall.

Evans, J. J., Garcia, E., Smith, M., Van Epps, A., Fosmire, M., & Matei, S. (2015). An assessment architecture for competency-based learning: Version 1.0. *Paper Presented at the Frontiers in Education Conference (FIE)*, El Paso, TX 2015. 32614 2015. IEEE.

Felder, R. M., Brent, R., & Prince, M. J. (2011). Engineering instructional development: Programs, best practices, and recommendations. *Journal of Engineering Education*, 100(1), 89–122.

Froyd, J. E., Wankat, P. C., & Smith, K. A. (2012). Five major shifts in 100 years of engineering education. *Proceedings of the IEEE, 100*(Special Centennial Issue), 1344–1360.

Gharaibeh, K., Harb, B., Salameh, H. B., Zoubi, A., Shamali, A., Murphy, N., & Brennan, C. (2013). Review and redesign of the curriculum of a masters program in telecommunications engineering –Towards an outcome-based approach. *European Journal of Engineering Education*, 38(2), 194–210.

Grodzicki, G., & Madigan, P. (2011). Outcomes-based assessment in instrumentation and measurement. *International Journal of Electrical Engineering Education*, 48(4), 451–462.

Hall, N., & Webb, D. (2014). Instructors' support of student autonomy in an introductory physics course. *Physical Review Special Topics-Physics Education Research*, 10(2), 020116-1–020116-22.

Henri, M., Johnson, M. D., & Nepal, B. (2017). A review of competency-based learning: Tools, assessments, and recommendations. *Journal of Engineering Education*, 106(4), 607–638.

Hsu, C.C., & Ho, C.-C. (2012). The design and implementation of a competency-based intelligent mobile learning system. *Expert Systems with Applications*, 39(9), 8030–8043.

Law, C. K. (1995). Using fuzzy numbers in educational grading system. *Fuzzy Set and Systems*, 83, 311–323.

Lopes, A. L. M., Lanzer, E. A., & Barcia, R. M. (1997). Fuzzy cross-evaluation of the performance of academic departments within a university. *Proceedings of the Canadian Institutional Research and Planning Association Conference*, Toronto, Canada, October, 19–21.

Ma, J., & Zhou, D. (2000). Fuzzy set approach to the assessment of student-centered learning. *IEEE Transactions on Education*, 43, 237–241.

McAuley, A., Stewart, B., Siemens, G., & Cormier, D. (2010). *The MOOC Model for Digital Practice*. Charlottetown: University of Prince Edward Island, 1–57.

Munoz-Guijosa, J. M., Bautista Paz, E., Verdú Ríos, M. F., Díaz Lantada, A., Lafont, P., Echávarri, J., Muñoz, J. L., Lorenzo, H., & Muñoz, J. (2009). Application of process re-engineering methods to enhance the teaching-learning process in a Mechanical Engineering Department. *International Journal of Engineering Education*, 25(1), 102–111.

O'Reilly, E. N. (2014). Correlations among perceived autonomy support, intrinsic motivation, and learning outcomes in an intensive foreign language program. *Theory and Practice in Language Studies*, 4(7), 1313–1318.

Pappano, L. (2012). The Year of the Mooc, *The New York Times*. Retrieved November 29, 2012.

Ratcliff, K. A., Arkin, R. M., & Dove, M. K. (1997). A method for assessing Undergraduate General Ed., The 37th AIR Annual Forum, Florida, May 18–21, 1997.

Rikakis, T., Tinapple, D., & Olson, L. (2013). The digital culture degree: A competency-based interdisciplinary program spanning engineering and the arts. *Paper presented at the Frontiers in Education Conference*, 2013 IEEE, Oklahoma City, US.

Ro, H. K., Merson, D., Lattuca, L. R., & Terenzini, P. T. (2015). Validity of the contextual competence scale for engineering students. *Journal of Engineering Education*, 104(1), 35–54.

Roe, E. A. (2015). Converting a traditional engineering technology program to a competency based, self-paces, open-entry/open-exit/format. *Paper presented at the ASEE Annual Conference and Exposition*, Seattle, WA.

Schmucker, K. J. (1984). *Fuzzy Sets, Natural Language Computations, and Risk Analysis*. Rockville, MD: Computer Science Press.

Sugeno, M., & Kang, G. T. (1988). Structure identification of fuzzy model. *Fuzzy Sets and Systems*, 28, 15–33.

Spelt, E. J. H., Luning, P. A., van Boekel, M. A. J. S., & Mulder, M. (2015). Constructively aligned teaching and learning in higher education in engineering: What do students perceive as contributing to the learning of interdisciplinary thinking? *European Journal of Engineering Education*, 40(5), 459–475.

Taskinen, P. H., Steimel, J., Gr€afe, L., Engell, S., & Frey, A. (2015). A competency model for process dynamics and control and its use for test construction at university level. *Peabody Journal of Education*, 90(4), 477–490.

Taylan, O., Ridwan, A., & Parsaei, H. (2017). Assessment of student learning by hybrid methods at program level, Engineering Education Letters, QScience, 1–14.

Tyler, R. W. (2013). *Basic Principles of Curriculum and Instruction*, Chicago, IL: University of Chicago Press.

UN General Assembly (2015). Universal Declaration of Human Rights, December 10, 1948, 217 A (III), available at: http://www.refworld.org/docid/3ae6b3712c.html.

Walther, J., Kellam, N., Sochacka, N., & Radcliffe, D. (2011). Engineering competence? An interpretive investigation of engineering students' professional formation. *Journal of Engineering Education*, 100(4), 703–740.

Woodrow, M., Bisby, L., & Torero, J. L. (2013). A nascent educational framework for fire safety engineering. *Fire Safety Journal*, 58, 180–194.
Zou, T. X. P., & Ko, E. I. (2012). Teamwork development across the curriculum for chemical engineering students in Hong Kong: Processes, outcomes and lessons learned. *Education for Chemical Engineers*, 7(3), 105–117.

chapter three

Safety engineering education truly helpful for human-centered engineering

Toward creation of mindset for bridging gap between engineers and users

Atsuo Murata
Okayama University

Contents

3.1 Introduction

It seems that engineering education is too much oriented to added value or usability creation of systems or products from the perspective of engineers. It is true that the added value is believed to increase sales and supports by potential users. The traditional safety engineering aimed at enhancing safety based on traditional human factors and ergonomics and reliability engineering approach, and seems to contribute to enhancing safety and reducing crashes, collisions, and disasters to some extent. Moreover, it supported engineering activities to manufacture products or systems that are safe and secure to use. In spite of such efforts to develop human-friendly products or systems, crashes, collisions, and disasters caused by an incompatible man–machine system or interface still don't decrease in both number and scale. This suggests that safety engineering education should be further improved and enhanced to address such issues.

As well as such a mission to aim at high usability and capacity of products or systems, it is also essential for engineers to develop a system or product that makes users feel safe and secure to use the system or product and can prevent users from being involved in an incident or accident while they are using the system or product. Generally speaking, we unconsciously tend to place more importance on the profit aspect of the product (develop usable products with high capacity to be supported by many users) than the safety of the product (Murata & Moriwaka, 2017). Murata and Moriwaka (2017) discussed why one cannot satisfy both economy and safety. An attempt was made to explain the collapse model of proper balance between safety and economy (efficiency or profit) induced by the following cognitive biases: (i) mental accounting, (ii) loss aversion, and (iii) discount of safety. Murata et al. (2015) and Murata (2017) also suggested that proper safety management requires us to take into account the cognitive biases (Bazerman et al., 2001, 2008, 2012; Kahneman, 2011) that lead to distorted decision-making and eventually induce critical crashes, collisions, or disasters. This indicates that the understanding of cognitive biases should be included in safety engineering education (discipline) or training so that such an education is truly helpful for human-centered engineering that makes user feel safe and secure for the products or systems. Realization of such education (discipline) or training needs to create a mindset that attempts to bridge gaps between engineers and users.

This chapter discusses how safety engineering education should be advanced. It is discussed how gaps between engineers and users, which potentially causes the creation of human–machine incompatible system with higher risks of unsafe behavior, should be compensated for in order to be truly helpful for human-centered engineering. In Section 3.2, the gap between engineers and users is explained using a concept of mental model (Genter et al., 1983), and the issue that stems from this gap is presented as an addressing issue in safety engineering education. The gap is also interpretable from the viewpoints of cognitive biases. In Section 3.3, the basic mechanism of cognitive biases is overviewed. In Section 3.1, the cognitive biases to which engineers are liable are demonstrated using a case study of Chinese Airline 140 crash. Section 3.5 discusses how safety engineering education should be advanced so that this can be truly helpful for humans (users) and become more human-centered. On the basis of the discussion, some proposals are made for advanced safety engineering education from the viewpoint of human factors engineering and irrational human mind (or decision-making). Section 3.6 summarizes how advanced safety engineering should be to bridge gaps between engineers and users, and how engineers' mindset should be fostered aiming at user-centered engineering that actually takes into account not only usability or capability (performance) but also safety.

3.2 Gap between engineers and users-mental model

There exists a gap of image for a system or product between engineers and users as shown in Figure 3.1. Figure 3.1 explains the concept of mental model (Genter et al., 1983). While users have their own model or image when they image to use a product or system, engineers have their own model or image for the development or design of a product or system. If the image of engineers coincides with that of users, the usability of system or product is judged to be satisfactory and usable. The unremitting and relentless efforts to repeat a usability test and bridge the gap between engineers and users are essential to compensate for the gap and realize such a coincidence. There are a lot of points in common between this and the concept of product or system liability shown in Figure 3.2.

Figure 3.2 shows the liability to a product or system from the viewpoints of both engineers and users. There is also a gap of liability between

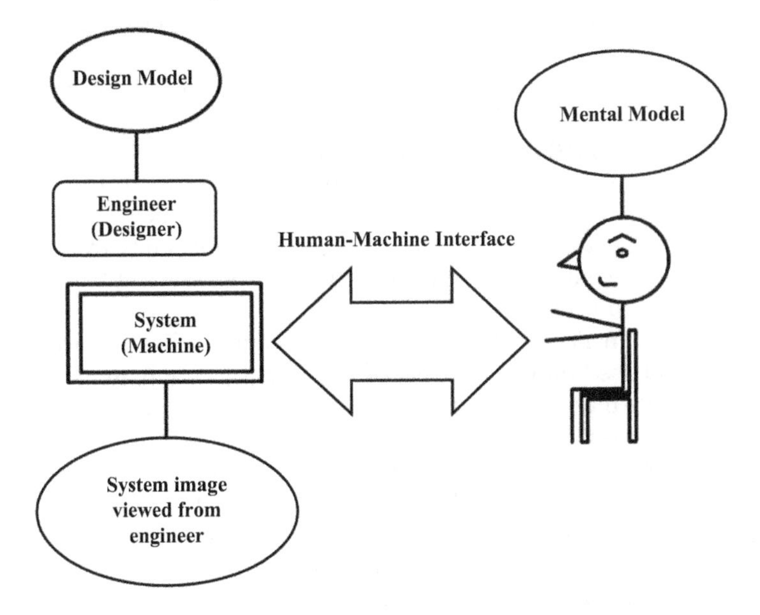

Figure 3.1 Concept of mental model.

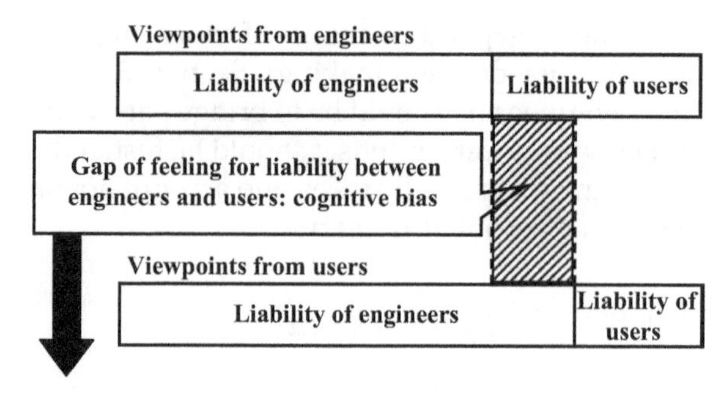

Figure 3.2 Gap of feeling for liability between engineers and users.

engineers and users. The liability of users viewed from engineers is larger than that viewed from users. This gap makes engineers misunderstand that they need not understand the traits or characteristics of users, and engineers eventually do not actively understand the mental model of users (what they need from a system or product). This hinders the development of system or product that is safe and secure to use for users, and leads to a lack in mindset to pursuit further safety persistently. In addition to the coincidence of mental model of engineers and users, engineers must exert themselves to bridge the gap of liability between engineers and users.

Although mentioned in more detail in Section 3.3, these gaps must come from human's cognitive biases. In other words, overconfidence or self-serving bias must be contributing to the gap. Self-serving bias forces us to attribute a success to one's own factors and a failure to other external factors. Engineers are overconfident in their ability to develop a product or system that never induces inconveniences or major safety troubles, and attribute such events (inconveniences or major safety troubles) to mistakes or errors of users.

As shown in Figure 3.3, there exist two approaches toward safety: an approach based on hazard (risk of incident) detection and that based on safety confirmation. The approach based on hazard (risk of incident) detection regards no detection of hazard (risk of incident) as normal, and does not attempt to actively confirm safety. There are many cases where the risk of incident exists in the system latently and damage the system not presently but later. Therefore, this is not enough to enhance safety. It goes without saying that the latter approach is more desirable than the former. Cognitive biases cause our irrational behavior, and it is possible that our distorted and irrational behavior becomes a trigger of critical

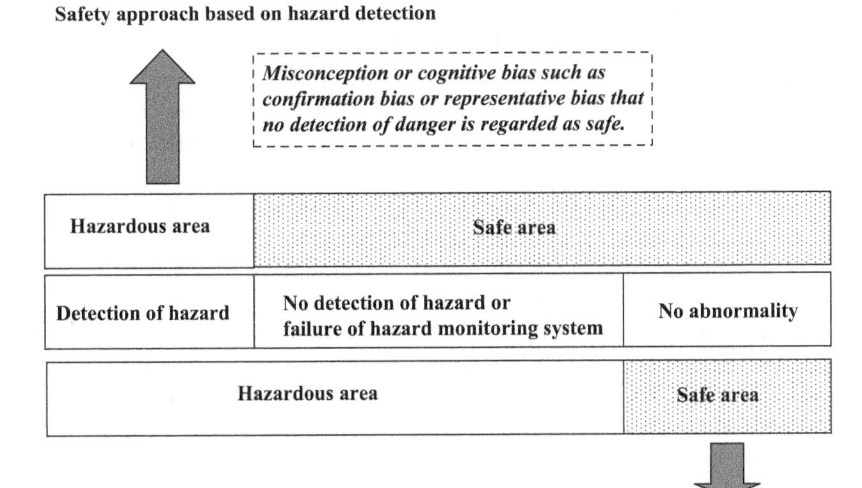

Figure 3.3 Two safety approaches: safety approach based on hazard (risk of incident) detection and that based on safety confirmation.

disasters, crashes, or collisions (Murata et al., 2015). Cognitive bias, such as availability bias, confirmation bias, and normalcy bias, makes us orient toward the safety approach based on hazard (risk of incident) detection. We usually can spend our daily life or jobs without experiencing danger, failure, or major incident. If such a state continues, we misunderstand that we can continue spending without such events. This leads to availability bias, confirmation bias, and normalcy bias (Dobelli, 2013). Based on the discussion, the safety approach based on the certain confirmation of safety is to be recommended.

3.3 Human's cognitive biases

Bazerman et al. (2001, 2008, and 2012) hypothesized that heuristics, such as availability, representativeness, confirmation or affect, cause biases, including confirmation biases, anchoring and adjustment, hindsight bias, availability bias, and conjunction fallacy. An event easily imaginable is more available than an event that is difficult to imagine. For example, the availability of the vividness of imaginable events biases our perception of the occurrence frequency of similar events. This might lead to wrong decision-making on the frequency of such events. Managers predict a salesperson's performance on the basis of an established category of salespeople, and do not make use of other information necessary for the accurate prediction. In other words, they tend to predict the future performance using only limited information. This corresponds to the representativeness heuristic. While this heuristic offers a proper approximation of salespeople in some cases, it can induce a biased understanding of salespeople and lead to serious errors in other cases. Such errors include ignorance of base rate or insensitivity to sample size pointed out by Kahneman (2011). People naturally tend to seek information that confirms their expectations and hypotheses, even though information disconfirming their expectations and hypotheses is actually more useful. This induces a biased recognition of causality and leads to serious errors. Figure 3.4 shows that not only heuristics but also overconfidence and framing are causes of biases (Murata & Yoshimura, 2015). Bounded awareness prevents one from focusing on useful, observable, and relevant information especially under uncertain situations. Due to such bounded awareness, it is valid to assume that we occasionally cannot behave rationally as is expected from standard economics. Moreover, it is reasonable to assume that our bounded awareness, under uncertain (and risky) situations, makes us more frequently experience cognitive biases (heuristic, overconfidence, and framing).

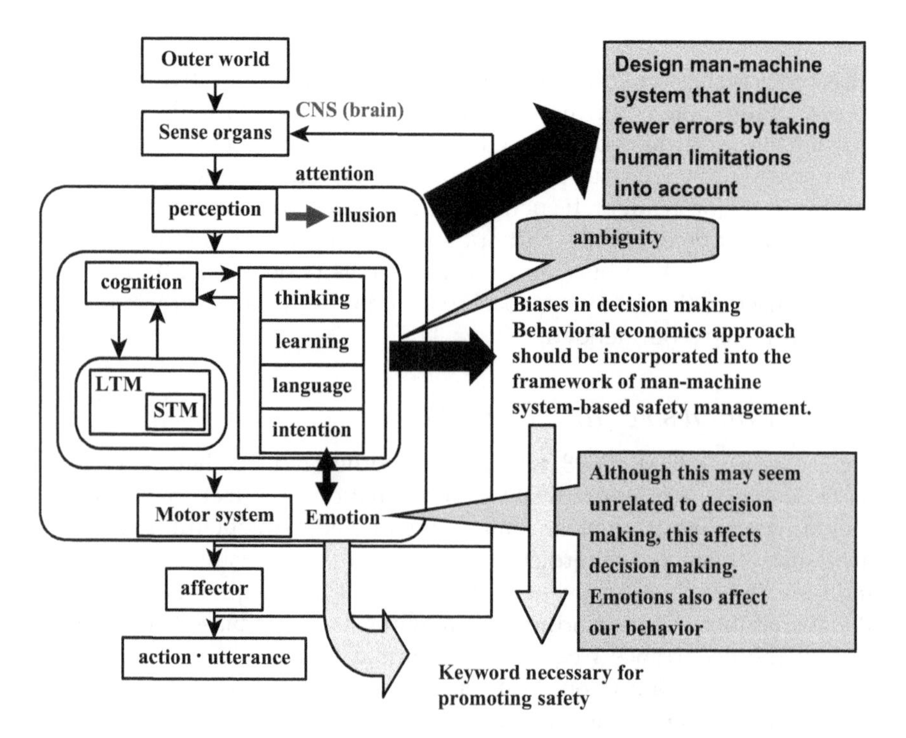

Figure 3.4 Each stage of cognitive information processing and function of cognitive bias and emotion.

We frequently tend to behave irrationally and are, in most cases, unaware of how and to what extent these irrational behaviors influence us. Such irrational tendencies definitely and unconsciously distort our decisions and, in the worst cases, lead to incidents, crashes, collisions, or disasters. Without the consideration of our bounded rationality (irrationality), we cannot analyze and remove a root cause of incidents, crashes, collisions, or disasters. Further analysis should be conducted on how cognitive biases distort decision-making, induce preconception, and become a trigger of incidents, crashes, collisions, or disasters. To this end, we must further clarify the mechanisms related to why we suffer from cognitive biases, under what conditions we are vulnerable to cognitive biases, what type of cognitive bias is potentially hazardous and readily leads to an unfavorable and unexpected incident, and when or how cognitive biases distort decision-making and become a trigger of errors, violations, and critical incidents.

Figure 3.4 shows human's cognitive information processing in which the basis of cognitive bias is included. The human error can occur at each process, that is, perception, memory (long- and short-term memory), and cognitive process such as judgment, creation, and decision-making. We suffer from illusions in the perception process, and long- and short-term memory is distorted in the storage and retrieval process. In this figure, the role of emotion is also depicted. The ambiguity in cognitions such as thinking, learning, intention, and decision-making is regarded to be the cause of cognitive biases. Emotion also plays a role in inducing cognitive biases. For example, owning some good induces a feeling of attachment, which leads to a cognitive bias called endowment effect. The endowment effect means that one tends to put a higher price when one owns it than when one doesn't own it.

It has been shown that how the cognitive biases lead to crashes, collisions, or disasters (Murata & Yoshimura, 2015). As shown in Murata et al. (2015), the cognitive biases should be paid more and more attention to so that the bias or distortion of decision-making should be avoided and eventually disasters, collisions, crashes of the system or product are effectively prevented. As mentioned in the next section using a case study of China Airlines 140 crash, in order to identify a root cause of the disaster, we should get further and detailed insight into the stage of cognitive information processing from which the cognitive biases stem and lead to such disasters.

Murata et al. (2015) have demonstrated how cognitive biases can be the root cause of incidents, crashes, collisions, or disasters throughout case studies. They emphasized the significance and criticality of systematically building the problem of cognitive biases into the framework of a man–machine system as shown in Figure 3.5. Moreover, they call attention to addressing human errors and preventing incidents, crashes,

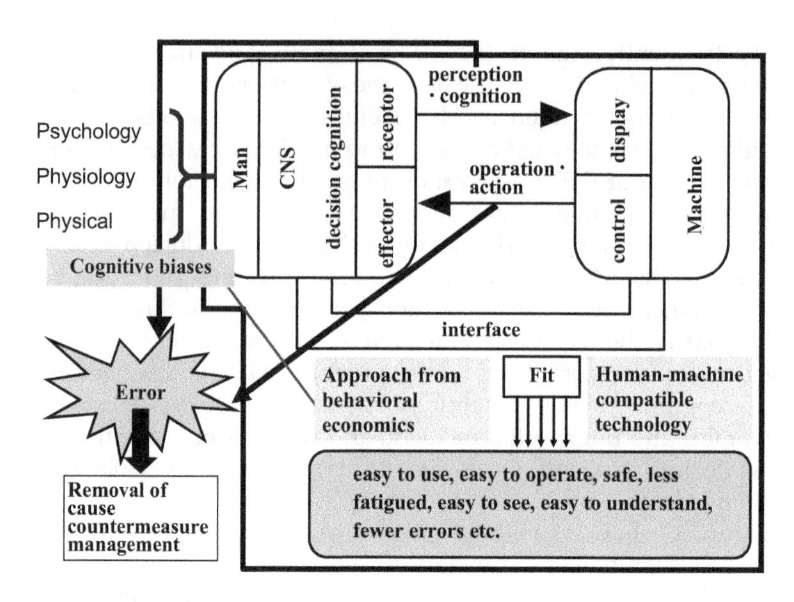

Figure 3.5 Framework of man–machine system that takes cognitive biases into account.

collisions, or disasters more effectively by considering cognitive biases. If designers, experts, or local or central officials of man–machine (man–society, man–economy or man–politics) systems do not understand the fallibility of humans, the limitation of human cognitive ability and the effect of emotion on behavior, the design of the systems could pose incompatibility issues, which, in turn, will induce critical errors or serious failures. To avoid such incompatibility, we must focus on when, why, and how cognitive biases overpower our way of thinking, distort it and lead us to make irrational decisions from the perspective of behavioral economics (Bazerman et al., 2001, 2008, 2012; Kahneman, 2011), as well as the traditional ergonomics and human factors approach. The design of man–machine (society, economy, or politics) compatibility (compatible technology) must be the key perspective for a preventive approach to human error-driven incidents, crashes, collisions, or disasters, and such an approach should be added to the framework of advanced safety engineering.

Gladwell (2008), referring to a concept of normal accident (Perrow, 1999, 2011), warned that a disaster, such as the NASA Challenger disaster, is unavoidable as long as we continue developing large-scale systems with high risks for the profit of humans. Such a situation corresponds to a vicious circle (repeated occurrences of similar critical incidents). Cognitive biases must leave such situations unresolved and, thus, hinder the progress of safety management and technology. A promising method to address such a problem might be to take into consideration and steadily eliminate cognitive biases that unexpectedly and unconsciously interfere with the functioning of large-scale systems, so that man–machine compatible systems can be established and maintained. Introducing appropriate safety interventions that ensure that cognitive biases do not eventually manifest themselves as causal factors of incidents, crashes, collisions, or disasters would enable one to address a cognitive-bias-related safety problem appropriately and to develop a man–machine compatible system (see also Figure 3.6). Advanced safety engineering education should actively incorporate such a subject into the education framework.

Figure 3.6 summarizes the concept of human–machine (society, economy, or politics) compatibility (compatible technology). Such compatibility will be helpful to avoid cognitive biases and lead to the decrease of disasters, collisions, crashes, or incidents. Advanced safety engineering education should also be oriented toward the design and development of human–machine compatible system free from cognitive biases. In order to realize this, the fallibility of human, the limitation of each process during cognitive information processing, the effect of emotion on behavior, and behavioral economics expertise, as well as traditional human factors and ergonomics approach, should be placed most importance on in advanced safety engineering education.

The next section deals with the gap between engineers and users in detail, which is attributed to cognitive biases. It is discussed how advanced safety engineering should work and be advanced to compensate for the gap between engineers and users.

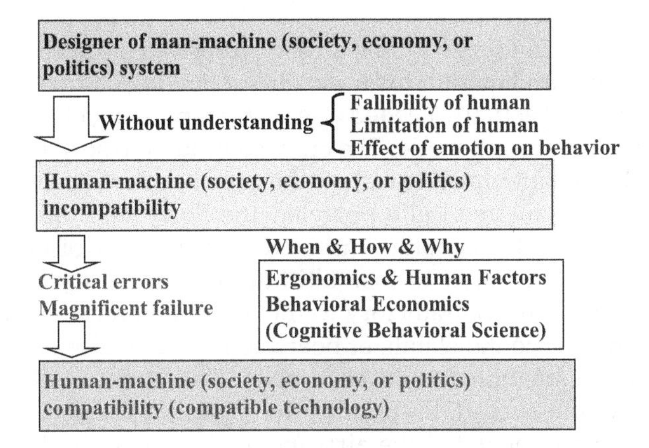

Figure 3.6 Enhanced human–machine compatibility that takes human's irrational characteristics into account.

3.4 *China Airlines 140 Crash caused by gap between engineers and users*

China Airlines Flight 140 was a regularly scheduled passenger flight from New Taiwan Taoyuan International Airport (Taipei, Taiwan) to Nagoya Airport (Nagoya, Japan). On 26 April 1994, the Airbus A300B4-622R was completing a routine flight and approaching to Nagoya Airport. Just before landing at Nagoya Airport, the crew made an error to inattentively press the takeoff/go-around (TO/GA) button. This button makes the throttle position to the same as takeoff and go-around. The direct cause of the crash was this human error and the subsequent failure to correct the control of aircraft as well as the airspeed (failure to recover from the error operation).

The flight took off from New Taiwan Taoyuan International Airport and was bound for Nagoya Airport. The descent of the flight started, and the airplane passed the outer marker. Just 5.6 km from the runway threshold at 1,000 ft, the airplane leveled off for about 15 s and continued descending until about 500 ft, where the aircraft was nose up in a steep climb due to the crew's error operation. Airspeed dropped quickly, and the airplane struck the ground.

There occurred a near miss (minor incident) with the Airbus A300-600R before China Airlines 140 crash. Airbus modified the air flight system so that the autopilot was disengaged when an input of certain manual control was applied on the control wheel in the go/around mode. This modification was not judged urgent by Airbus. The aircraft that had crashed had been scheduled to receive the upgrade. The aircraft had not received the upgrade immediately after the near miss happened in Helsinki, because China Airlines announced that the modifications were not essential and urgent.

The crews misunderstood that strongly pressing a control stick could call off the go-around mode. This was right for Bowing's aircraft. However, Airbus A300-600R actually needed a sequential pressing of a few buttons to call off the go-around mode. The crew continued to strongly press the control stick, and attempted to call off the go-around mode. However, the go-around mode automatically operated the horizontal tail and forced the airplane to rise. The captain continued to push a control stick strongly and repeatedly attempted to lose altitude to prepare for the landing. These operations are contradictory to the aircraft, because the computer controlled the horizontal stabilizer board to raise the nose of the aircraft and the captain attempted to lower its nose. The captain could not lower the altitude of the aircraft. Eventually, the captain made decision to try a landing again and operated the airplane to rise (go-around) again. The force to raise the nose had been competing with that to lower the nose for a while. The force to lower the nose was weakened suddenly due to the continuation of such contradictory operations. The nose suddenly rose, and the aircraft became vertical. The aircraft inevitably lost buoyant and was thrown to the ground.

The direct cause of the crash was that the crew could not cancel the go-around mode. There are few opportunities to call off the go-around mode during the actual flight experiences, and the crew was not accustomed to this operation. It is no use to be confused about the functions or operations that are rarely used. The design and development of a system itself, the function or operation of which even an operator beyond the average competence cannot fully understand, is problematic. The safety engineering education should bear this in mind.

The replacement of error-prone operations or tasks by automation or computerization is one of the effective countermeasures to prevent crashes caused by human errors. It is very important to replace tasks or jobs that frequently cause human errors by a computerized or automated machine and exclude such factors in the process of designing systems. Japanese Tokaido-Sanyo bullet train (Shinkansen) has not been experiencing collisions or crashes for about 50 years since its opening of business. This is because the bullet train system did not introduce railroad crossings the probability of collisions or crashes of which is the highest. Airbus attempted to promote automation of their aircraft so that error-prone human factors can be excluded as much as possible, because major causes of crashes are human errors or mistakes. It goes without saying that this should be recommended. Even how the automated system is advanced, the interface between man and machine is sure to exist. The interface is the most vulnerable of all system components. Which should take an initiative for the safe navigation of aircraft, pilots, or automated and computerized systems? While we frequently commit errors, we excel in the ability to receive a lot of information and make judgment in an instant. Therefore, the answer is self-evident, and an initiative should be left to humans. The computerized and automated cockpit system of the aircraft continued to operate against the direction by the crew members. Such an event should be avoided so that critical crashes caused by a human error should not occur.

The important lesson was that the system should be robust so that the mistakenly selected go-around mode can be readily switched to the manual mode. The initiative of the aircraft's cockpit should be definitely left to the pilot. However, it must also be noted that another type of errors such as the failure of landing potentially occur even in the manual-mode operation.

One of factors behind the crash was that Airbus did not learn or extract a lesson from a similar trouble (near misses (minor incident)). A similar trouble of mistakenly (inattentively) pressing the takeoff/go-around (TO/GA) button occurred at Helsinki International Airport,

Finland 5 years before the crash mentioned above. Then, the crew managed to make use of the operation manual and call off the go-around mode. Airbus did not appropriately correspond to this near miss (minor incident), and they did not take this serious and emergent and force to alert the undo function so that the crew can call off the turn-around mode easily like the system used in Bowing (pressing a control stick). Moreover, Airbus did not regard this alteration mandatory (essential), and recommended that this alteration should be charged if each airline required Airbus to alert the undo function above. They were apparently lacking in the mindset to learn the importance of easy-to-understand undo function of go-around mode from the near miss (minor incident) occurred at Helsinki International Airport. The pilot, due to the effect of habituation and normalcy bias, thinks that their error operation can be called off (undone) readily by pressing a control stick. They are usually educated (trained) on Bowing Aircraft. The engineer must understand such characteristics of pilots and recognize the gap between engineers and pilots, as shown in Figures 3.1–3.3.

Without such understanding, engineers cannot design and develop an aircraft that is truly usable from the viewpoints of both usability and safety. The recommendation in this chapter is, like relentless and sustainable usability test of product, that the pilots should participate in the development (especially the design of aircraft cockpit) from the beginning. This enables engineers to develop an aircraft that is truly helpful for pilots. Orienting at human-centered engineering needs participation of all people involved in the system. As shown in this case study, there apparently existed a gap between engineers and pilots. The design of systems, in particular, large-scale systems should be carried out jointly and cooperatively by engineers and users (pilots in this case). Using the concepts of Figures 3.1–3.3, the disaster can be summarized as shown in Figures 3.7–3.9.

The gap between engineers and users in the case of China Airlines 140 crash is summarized in Figure 3.7. Engineers were lacking in system image that considered pilots' characteristics or opinion. There were many psychological aspects that engineers could not understand. Pilots were not used to an automated and computerized system, and misunderstand that the go-around mode could be called off by one pressing of control stick according to their trait. Engineers did not assume that pilots are affected by their prejudice and do not understand completely the automated system. Therefore, the gap apparently existed between engineers and pilots, which caused this crash. To suppress the disagreement between engineers and pilots to a minimum, engineers should understand the characteristics or traits of pilot during a flight, and the development by the participation of both engineers and pilots should be promoted. In other words, such an approach should be actively adopted for the development and design of

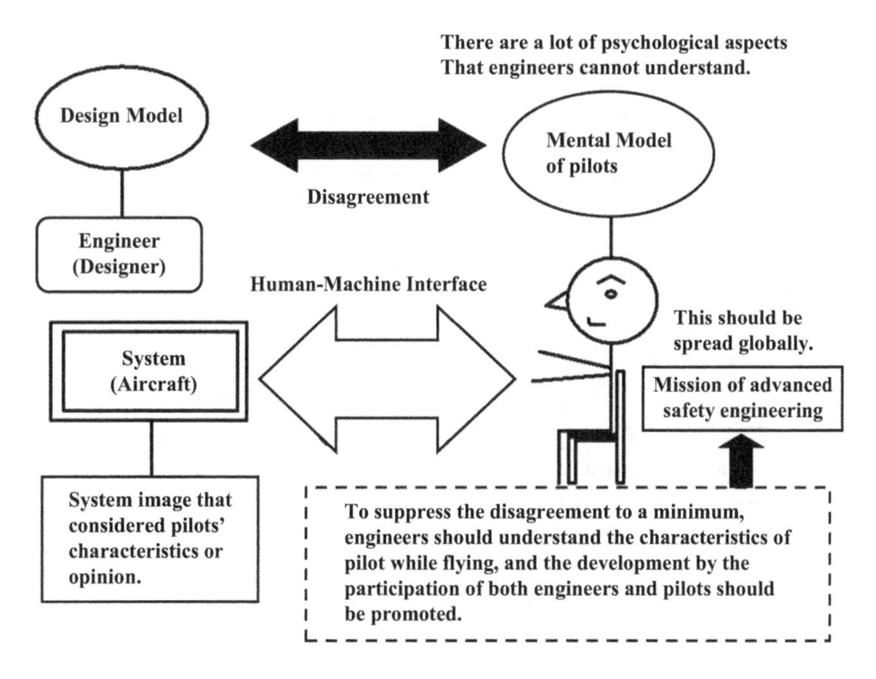

Figure 3.7 Application of China Airlines 140 crash to mental model concept (Figure 3.1).

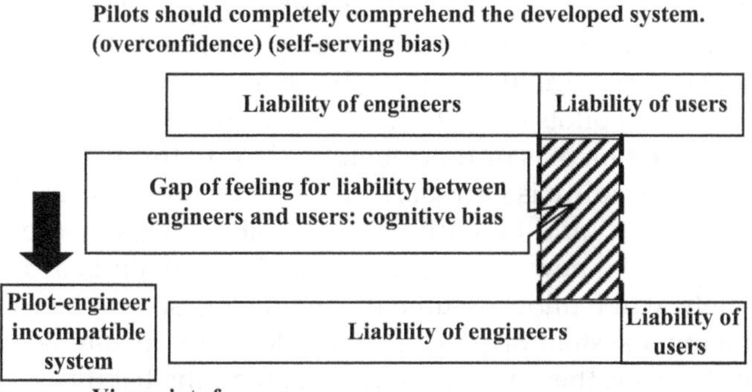

Figure 3.8 Application of China Airlines 140 crash to liability gap model (Figure 3.2).

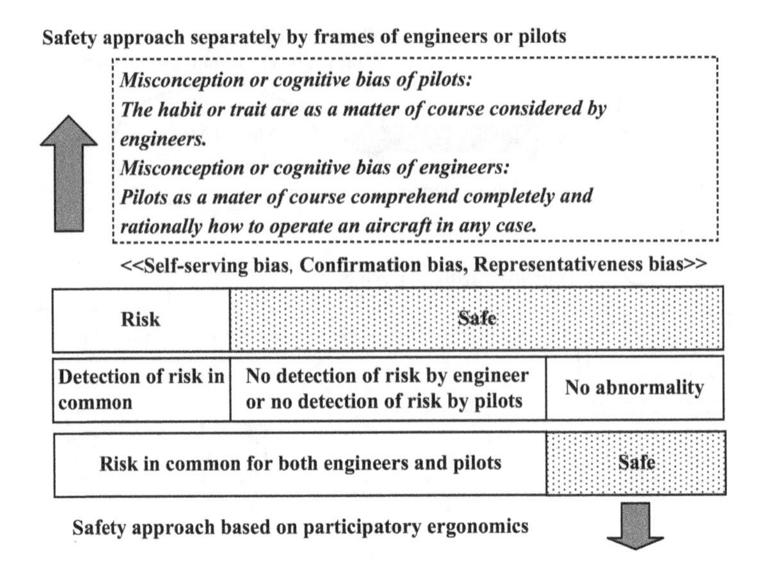
Safety approach separately by frames of engineers or pilots
Misconception or cognitive bias of pilots:
The habit or trait are as a matter of course considered by engineers.
Misconception or cognitive bias of engineers:
Pilots as a mater of course comprehend completely and rationally how to operate an aircraft in any case.
<<Self-serving bias, Confirmation bias, Representativeness bias>>
Risk
Safe
Detection of risk in common
No detection of risk by engineer or no detection of risk by pilots
No abnormality
Risk in common for both engineers and pilots
Safe
Safety approach based on participatory ergonomics

Figure 3.9 Application of China Airlines 140 crash to safety approach separately by frames of engineers and pilots (Figure 3.3).

systems or products so that the system or product is tolerant to human errors and further enhance safety.

Figure 3.8 applies this case study to the model in Figure 3.2. Viewed or framed from engineers, pilots should actively learn from the operation manual and completely comprehend the developed system. This corresponds to overconfidence to their technology and self-serving bias to think that it is the fault of pilots if they neglect to learn completely from an operation manual. On the other hand, viewed or framed from users, even skilled pilots cannot completely refrain from cognitive biases such as confirmation bias or representative bias to think mistakenly that their routine procedure is necessarily reflected in any system. Engineers think that pilots should be entirely responsible for the comprehension of how to operate a new system as long as they are on board to fly the aircraft safely, while pilots assume that engineers should be responsible for developing a system or product that pilots can operate without confusion. In this way, there is apparently a gap on the liability of a system or product between engineers and pilots, and this causes incidents or crashes as mentioned in this section. To compensate for this gap, pilot-oriented safety engineering approach is indispensable, and the system or product should be designed and developed by the participation of all people involved.

Figure 3.9 applies this case to the model in Figure 3.3. Safety approach is conducted separately by frames of engineers and pilots as shown in the upper part of Figure 3.9. Only the risk recognized in common by both engineers and pilots is regarded as a risk, and the risk separately recognized by either group (engineers or pilots) is not recognized as a risk. This must be caused by the gap illustrated in Figures 3.7 and 3.8. Misconception or cognitive bias of pilots makes themselves think that their habits or traits are as a matter of course considered by engineers. Misconception or cognitive bias of engineers makes themselves assume that pilots as a matter of course comprehend completely and rationally how to operate an aircraft in any case. Self-serving bias, confirmation bias, and representativeness bias must work in such a case and make both engineers and pilots misunderstand that only the risk in common should be paid attention to and the risk recognized separately by engineers and pilots is not regarded as important. Thus, the engineers of Airbus, during the development process, must have assumed that their automated and computerized systems can be as a matter of course understood appropriately by the pilots. Safety approach based on participatory ergonomics must be effective to compensate for such a difference in recognition as mentioned in Figures 3.7 and 3.8.

The gap of recognition of safety between engineers and pilots can be expressed in three different manners, as shown in Figures 3.7–3.9. The most important thing for advanced safety engineering education is a mindset to compensate for this gap.

3.5 *Safety engineering education truly helpful for humans and some proposals from human factors engineering*

Figure 3.10 summarizes how the framework of traditional man–machine systems should be improved and incorporated into advanced safety engineering education. Engineers are apparently lack in consideration of human's cognitive bias that gives rise to a gap between engineers (designers) of system or product and their users. Based on the traditional framework of man–machine system according to the principle of "fitting the task to the man," the safety engineering education has been engaging in a variety of educations such as reliability engineering and analysis, failure diagnosis, safety design, and mechanism of hardware, safety management strategy, human error management (detection, analysis, and recovery from human error), human factors and ergonomics, and industrial psychology. As such a framework itself systematically integrates a variety of disciplines, it goes without saying that this framework to some extent contributed to advancing the safety engineering education. However, as mentioned in Sections 3.2–3.4, such an approach apparently lacks in efforts to compensate for the gap between engineers and users, and should be

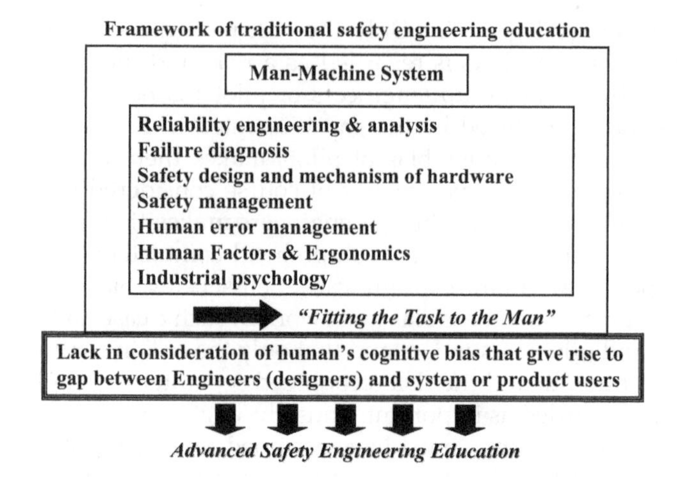

Figure 3.10 Framework of traditional safety engineering education and its lack in consideration of human's cognitive biases.

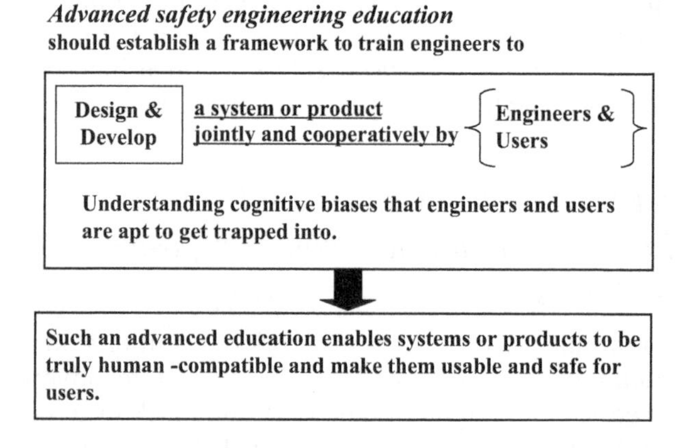

Figure 3.11 Summary of how advanced safety engineering should be.

further advanced so that the gap can be definitely recognized and compensated for by the involvement of all members related to the system or product and using an approach such as participatory ergonomics.

Figure 3.11 summarizes the framework of advanced safety engineering education. An advanced safety engineer should train and educate engineers to design and develop a system or product jointly, cooperatively, and bilaterally by the participation of engineers and users. Engineers should also be trained to understand cognitive biases that engineers and users are apt to get trapped into, and take into account such properties in designing and developing a system or product. The author believes that such an advanced safety engineering education framework enables systems or products to be truly human-compatible and make them usable and eventually safe and secure for users.

3.6 Summary: Toward creation of engineers' mindset essential for bridging gap between engineers and users-oriented to not only usability but also safety

The concept and framework of safety engineering approach mentioned in this chapter is indispensable for human-centered engineering. Participation of not only engineers but also users eventually makes the development effective and safe and secure to use, and leads to error-free or error-robust systems or products. Educating and training engineers such a mindset will be sure to contribute to the further advances of safety engineering education. As the concept of ergonomics shows that the task should be fit to the man, this concept of fitting the task to the man should be practiced not only by engineers but also by users. This leads to the concept of participatory ergonomics, which should be adopted in advanced safety engineering all over the world (globally) so that the safety and secure of man–machine compatible systems is further enhanced.

References

Amano, Y. *The Fukushima Daiichi Accident.* Report by the Director General, International Atomic Energy Agency: Vienna, Austria, 2015.

Bazerman, M.H., and Moore, D.A. *Judgment in Managerial Decision Making,* Harvard University Press: Cambridge, MA, 2001.

Bazerman, M.H., and Watkins, M.D. *Predictable Surprises,* Harvard Business School Press: Cambridge, MA, 2008.

Bazerman, M.H., and Tenbrunsel, A.E. *Blind Spots: Why We Fail to Do What's Right and What to Do About It,* Princeton University Press: Princeton, NJ, 2012.

Culati, R., Casto, C., and Krontiris, C. How the Other Fukushima Plant Survived, *Harvard Business Review,* August 2014.

Dekker, S. *The Field Guide to Understanding Human Error,* Ashgate Publishing: Burlington, VT, 2006.

Dobelli, R. *The Art of Thinking Clearly,* Harper: New York, 2013.

Gardner, D. *Risk—The Science and Politics of Fear,* McClelland & Stewart: Toronto, Canada, 2008.

Gardner, D. *Future Babble—Why Expert Prediction Fall and Why We Believe Them Anyway,* Emblem: Toronto, Canada, 2010.

Genter, D., and Stevens, A.L. *Mental Models,* Lawrence Erlbaum Associates Publishers: Hillsdale, NJ, 1983.

Gladwell, M. The ethnic theory of plane crash. In Gladwell, M. (ed.), *Outliers, Back Bay Books,* pp. 206–261, 2008.

Gladwell, M. Blowup. In *What the Dog Saw*, pp. 345–358, 2009.

Helmreich, R.L. Anatomy of a system accident: The crash of Avianca flight 052. *International Journal of Aviation Psychology*, 4(3), 265–284, 1994.

Helmreich, R.L., and Merritt, A. Culture in the cockpit: Do Hofstede's dimensions replicate? *Journal of Cross-Cultural Psychology*, 31(3), 283–301, 2000.

Hollnagel, E., Woods, D.D., and Leveson, N. (Eds.) *Resilience Engineering: Concepts and Precepts*, Ashgate: Aldershot, UK, 2006.

Hollangel, E., and Fujita, Y. The Fukushima Disaster—Systemic failures as the lack of resilience. *Nuclear Engineering and Technology*, 45(1), 13–20, 2013.

Hollnagel, E., Paries, J., Woods, D. D., and Wreathall, J. (Eds.) *Resilience Engineering in Practice: A Guidebook*, Ashgate: Farnham, UK, 2011.

Kahneman, D. *Thinking, Fast and Slow*, Penguin Books: London, UK, 2011.

Meshkait, N., and Khase, Y. Operator's improvisation in complex technological systems: Successfully tackling ambiguity, enhancing resiliency and the last resort to averting disaster. *Journal of Contingencies and Crisis Management*, 23(2), 90–96, 2015.

Mullainathan, S., and Shafir, E. *Scarcity: The New Science of Having Less and How It Defines Our Lives*. Picador: New York, 2014.

Murata, A., and Yoshimura, H. Statistics of a variety of cognitive biases in decision making in crucial accident analyses. *Proceedings of 6th International Conference on Applied Human Factors and Ergonomics (AHFE 2015) and the Affiliated Conferences*, AHFE 2015, Volume 3, pp. 3898–3905, 2015.

Murata, A., Nakamura, T., and Karwowski, W. Influence of cognitive biases in distorting decision making and leading to critical unfavorable incidents. *Safety*, 1(1), 44–58, 2015.

Murata, A. Cultural influences on cognitive biases in judgment and decision making: On the need for new theory and models for accidents and safety. In *Modeling Sociocultural Influences on Decision Making – Understanding Conflict, Enabling Stability*; J.V. Cohn, S. Schatz, H. Freeman, and D.J.Y. Combs (Eds.), CRC Press: Boca Raton, FL, pp. 103–109, 2017.

Murata, A., and Moriwaka, M. Anomaly in safety management: Is it constantly possible to make safety compatible with economy? In *Advances in Safety Management and Human Factors* (Advances in Intelligent Systems and Computing 604); P. Arezes (Ed.), Springer, Berlin, pp. 45–54, 2017.

Nakamura, A., and Kikuchi, M. What we know, and what we have not learned: Triple disasters and the Fukushima nuclear fiasco in Japan. *Public Administration Review*, 71, 893–899, 2011.

National Research Council of the National Academy. *Lessons Learned from the Fukushima Nuclear Accident for Improving Safety of US Nuclear Plants*, National Academies Press: Washington, DC, 2014.

Perrow, C. *Normal Accidents: Living with High-Risk Technologies*. Princeton University Press: Princeton, NJ, 1999.

Perrow, C. *The Next Catastrophe: Reducing Our Vulnerabilities to Natural, Industrial, and Terrorist Disasters*, Princeton University Press: Princeton, NJ, 2011.

Perrow, C. Fukushima and the inevitability of accidents. *Bulletin of the Atomic Scientists*, 67(6), 44–52, 2011.

Perrow, C. Fukushima, risk and probability: Expect the unexpected. *Bulletin of the Atomic Scientists*, 67(4), 44–52, 2011.

Perrow, C. Nuclear Denial: From Hiroshima to Fukushima. *Bulletin of the Atomic Scientists*, 69(5), 56–67, 2013.

Pillay, M. Advancing organisational health and safety management: Are we learning the right lessons? In *Advances in Safety Management and Human Factors* (Advances in Intelligent Systems and Computing 604); P. Arezes (Ed.), Springer: Cham, Switzerland, pp. 37–44, 2017.
Reason, J. *Managing the Risks of Organizational Accidents Revisited*, Ashgate Publishing: Surrey, UK, 1997.
Reason, J. *Organizational Accidents Revisited*, Ashgate Publishing: Surrey, UK, 2016.
Suzuki, A. Managing the Fukushima challenge. *Risk Analysis*, 34(7), 1240–1256, 2014.
Syed, M. *Black Box Thinking: Marginal Gains and the Secrets of High Performance*, John Murray Publishers Ltd: New York, 2016.
The National Diet of Japan. The official report of the Fukushima Nuclear Accident Independent Investigation Commission, 2012.

chapter four

Globalization issues and their impacts on engineering education

Radhey Sharma
West Virginia University

Contents

4.1 Introduction

The interaction of humanity with its habitat is essential for food, shelter, security, safety, health, energy, mobility, and human progression and growth. Engineering in one form or the other is needed in the multifaceted human endeavor. Engineering is as old as humanity itself, and it has played a critical role in development of human civilization. There are examples of engineering and its use in well-being and progress of humanity around the world including in the ancient civilizations such as Indian, Chinese, Egyptian, Persian, and Roman.

Engineering techniques and skills are useable anywhere with appropriate local adaptations. In ancient times, human mobility was limited and global connectivity did not exist. Most of the interactions were local and modes of travel were slow. This led to development of engineering discipline in different parts of the world with hardly any exchange of ideas and engineering know how among different regions of the world.

4.2 Globalization and global issues

The human quest has been to enhance mobility and connectivity. However, this was very limited and slow until the 18th century. Global interactions became feasible with the advent of modern means of transport such as air travel. As the air travel became more common and available to a broader cross-section of population, human interactions at global level accelerated. Connectivity was still limited and slow. Internet transformed connectivity and communications.

Pace and scope of globalization accelerated with new technology enabling mobility and connectivity. In the 21st century, technological transformation and global connectivity impacted entire human endeavor including engineering education. Global demand for high-quality engineering education is increasing exponentially with rising global population. Some of the important features of this demand for engineering education are as follows:

- Engineering work is becoming more globalized with distributed manufacturing. It is common to have production of different parts of a product at different global sites and then assembling and distributing it in different regions of global market. Some examples are the Apple iPhone and auto manufacturing. These need engineering professionals who can work at different global sites as well as interact with diverse global engineering professionals.
- The need for high-quality engineering education is increasing at all levels—undergraduate, graduate, and professional development and training.
- The student population is diverse which includes students starting from freshman, transfer, working professionals who want to enhance their engineering education and skills.
- Increasing number of students opt for education abroad experience.
- Students want to undertake internships and many prefer these internships at global locations so as to gain cultural exposure and enhance their employability.
- Higher education including engineering education is costly and availability of high-quality education is still very limited in many parts of the world.
- Engineering is a tool for economic development. Many parts of the world still lag in economic development and need a well-trained high-quality engineering work force. Additionally, centers of engineering education (e.g., universities) have significant economic impact and contribute to the well-being of the local population.
- There is greater consciousness and need for sustainable development and care of the environment. Issues of global warming and

environmental pollution need innovative engineering solutions and new engineering graduates are expected to handle these issues.
- Emerging global issues such as energy security, high-quality and affordable healthcare, water security, cybersecurity, cyber–physical infrastructure, smart cities, and food security need innovative solutions with interdisciplinary collaborations and global partnerships. New engineers are expected to have global perspective, skills for multidisciplinary collaborations, effective in global teams of professionals, and critical thinking skills. US National Academy of Engineering (NAE) provided an overview of grand challenges and the need to address these challenges (see NAE report 2008 and updated in 2017).

4.3 Engineering education

Globalization and issues emanating from it are increasingly becoming important. The pace of technological transformation is further accelerating the scope and reach of globalization. Engineering education is primarily provided by the institutions of higher education (e.g., universities). Universities intrinsically are slow in adapting to change. This creates a challenge as the market place is changing at a rapid pace whereas the universities are not used to such fast pace of change. Engineering education is provided by the universities and the strategies and tools for changing and adapting to the market place and to the expectations of new students and lifelong learners need to be considered and addressed.

Engineering curriculum need to be reviewed and revised to cater to the changing needs and expectations of the new students as well as the global engineering market place. For example, some US engineering schools have reduced the credit hours needed for an engineering degree. However, such changes are more driven by the local factors rather than collective thinking with global perspective. New curriculums need to address the needs and expectations of future students in a globalized world.

Along with the changes in the curriculum, there is need for standardization of quality measures. For example, accreditation of engineering programs by the Accreditation Board for Engineering and Technology (ABET) is common practice in the US engineering schools and recently some other countries are also adopting this practice. However, there is need to have a global standard wherein all global stakeholders are consulted so that the quality standards could be simplified and useful. For example, it should become easy for a student to study abroad and transfer the credits to home institution or to future educational institution where the student might go for subsequent education and training. It is not uncommon that students struggle in transferring their credits from their education abroad courses.

Demand for education abroad and global internships is increasing as the employers expect the new engineers to be effective in global market place. There are variations in content, availability, emphasis, quality, and financial support for education aboard programs offered by the universities. This is an area, which needs a significant improvements and transformation with participation of global stakeholders.

It is critical that besides graduate students, undergraduate students should have opportunities and access to cutting-edge research. It is important for two simple reasons. First, when students are exposed to high-quality and exciting research, they might decide to pursue graduate studies, which in turn will benefit the students as well as the host institutions. Second, even if some of the students decide not to pursue graduate studies, they will go from the university better informed with new horizons and would be better equipped to excel in industry and can better compete and collaborate at national and global level. The author, as Department Chairman, led this positive change for the Department of Civil and Environmental Engineering at the West Virginia University for undergraduate research experience and started a new Summer Undergraduate Research Fellowship (SURF) program in 2011. This SURF initiative was highly successful in providing useful experiential learning experience to undergraduate engineering students (see details in Sharma and Unnikrishnan, 2018). Some of the salient features of the SURF program included financial support for the students, all departmental faculty who participated in the program, any student with a grade point average of 3.0 or higher who could participate in the program, and research projects that were decided through student and individual faculty who served as research advisor for the student. Many of these students were excited about research and decided to pursue graduate studies. Also, several students won prestigious graduate fellowships from funding agencies such as National Science Foundation (NSF). A number of these SURF students published their research in conference proceedings and journals. Based on the feedback from industry, the students who went to industry after SURF experience were asset to their organizations in critical thinking skills as well as taking on lifelong learning responsibilities. Here it would be appropriate to mention that many universities have breadth and depth in research activities and undergraduate students can immensely benefit from these research opportunities. It would, however, be unquestionably essential to have in place proactive plans to further expand and enhance research enterprise so that undergraduate students can benefit. It is important to provide pathways to undergraduate and graduate students to undertake research at global site also so that the students have opportunities to develop global perspective and experience.

Internships provide new perspective and valuable experience to students and the author had been involved in promoting such opportunities. The author worked in close cooperation with the Director of Engineering College Corporate Relations at WVU to expand and enhance opportunities for internships. Additionally, interactions and nurturing relations with alumni and supporting industries were useful in increasing internship opportunities for students. A similar approach was developed for international internships by the author during his tenure as Associate Vice President for Global Strategies and International Affairs.

It is also important to recognize that student learning is a continuum and significant learning takes place outside the formal classroom. In this context, the author would like to point out experience that was gained while serving as Resident Faculty Leader for student affairs for four halls of residence at the University of Bradford, England. It was an exceptional experience in promoting and fostering an institutional atmosphere of holistic student learning with synergy of students' curricular and co-curricular experiences to support academic and personal success.

The above example can serve as template for engineering educators to develop opportunities for their students for experiential learning, internships, and education abroad.

Accessibility and affordability are vital challenges that higher education institutions face and need to develop specific strategies and plans to address these issues. This varies from institution to institution depending on the mission, resources, and flexibility. These issues and their impact vary even more from country to country. However, with the technological transformation and connectivity, some of these issues are being addressed through online and distance education. Nonetheless, this is an area where there is need for significant progress with global reach. Online and distance education could further be enhanced by universities in partnerships with entertainment industry.

4.4 *Influence of globalization on engineering education and research*

Globalization is still spreading in its reach with digital technology and its availability to wider global population. Digital technology and various platforms for social media are still evolving and these have a profound impact on education including engineering education. It is interesting to note that some countries which did not have major infrastructure for connectivity are able to leap frog by directly adopting the latest technology in the areas of Internet, digital, mobile communication devices, etc. An example of digital technology and social media as tool for engineering research is provided by Tang et al. (2017).

4.5 Thoughts on future developments in engineering education in the environment of globalization

The institutions of higher education (e.g., universities in the US) are adapting to needs of new age students, developing new financial models to make education accessible and affordable, and adapting the curriculum so that the needs of employers be met. In the US, private institutions are making significant contributions in transforming online education. Further development of online delivery is important and issues relating to quality of online and face-to-face education will become increasingly important in globalized world. There is need for global forums for exchange of ideas and sharing of experiences by diverse stakeholders. A large number of lifelong learners are utilizing the resources available for enhancing their skills. The open access and free content available for lifelong learners is becoming available in the field of engineering.

4.6 Conclusions

In last couple of decades, the pace of globalization has accelerated with technological transformation and increasingly interconnected and interdependent world. In this global ecosystem, engineering work is becoming globalized such as distributed manufacturing. It is common to have production of different parts of a product at different global sites and then assembling and distributing it in different regions of global market. Some examples are the Apple iPhone and auto manufacturing. This needs engineering professionals who can work at different global sites as well as interact with diverse global engineering professionals. Global challenges such as energy security, high-quality and affordable healthcare, water security, cybersecurity, cyber–physical infrastructure, smart cities, and food security need innovative solutions with interdisciplinary collaborations and global partnerships. New engineers are expected to have global perspective, skills for multidisciplinary collaborations, effective in global teams of professionals, and critical thinking skills. Globalization and the issues arising out of it are impacting the engineering education also. Additionally, new students use digital technology, and learning with the aid of digital devices is becoming common. Universities are still adapting to cater to demands and issues arising out of globalization. There are various aspects such curriculum, common standards, delivery systems, experiential learning, education abroad, internships at global sites, and global collaboration and partnerships are changing in nature and scope and will need to be addressed in more comprehensive manner at global level. New technology and innovative methods will need to be used to make engineering education accessible and affordable in various parts of the world. Overall, globalization is exciting for engineering education and with proper strategic planning and shared approach it will help in global economic development and well-being of the communities around the world.

References

NAE Report on NAE Grand Challenges for Engineering (2008 and updated in 2017). http://www.engineeringchallenges.org/File.aspx?id=11574&v=34765dff.

Sharma, R.S., and Unnikrishnan, A. (2018). Evaluating student perception of civil and environmental engineering undergraduate research. *Transportation Research Board*, Washington, DC.

Tang, L., Zhang. Y., Dai, F., Yoon, Y., Song, Y., and Sharma, R.S. (2017). Social Media Data Analytics for the U.S. Construction Industry: A Preliminary Study on Twitter. *ASCE Journal of Management in Engineering*, 33(6): 04017038.

chapter five

How to use interdisciplinary team innovation to foster crucial engineering competencies?

Mona Enell-Nilsson, Minna-Maarit Jaskari, and Jussi Kantola
University of Vaasa

Contents

5.1 Introduction

Globalization, technological development, and changing demographics are among megatrends that keep the world changing. For example, internet of things, big data, data analytics, and artificial intelligence are a source of market disruptions. The ongoing changes affect the working life as well as requirements for jobs, people, and machines. The continuous change in these areas result in changes in the requirements for engineering work, i.e., how engineering work should be organized, and what the new work roles and their requirements are. Thus, the future workplace requires new roles with a set of new skills. Future engineers tackle problems with high complexity and interdisciplinarity. Thus, competencies such as complex problem-solving, teamwork, and effective communication are important for future engineers (Passow & Passow, 2017). This is an important signal also for engineering educators. How can we teach these crucial competencies already during time spent at the university? How do we create a learning environment where students can practice their skills in a safe error-and-trial environment where experimenting is allowed and wished for in order to meet the challenges faced after graduation?

We are not alone with these thoughts, as one of the most active research areas in engineering education is competencies (Henri et al., 2017). Competencies and especially differentiating competencies are directly linked to superior performance at work (Boyatzis, 1982, 2008; Spencer & Spencer, 1993)—engineering work in this context. As competencies link engineers to good and superior performance, it is crucial for engineering educators to understand how to teach and support the development of competencies already during education. Thus, engineering competence development forms a very interesting, contemporary, and valuable research area.

In this chapter, we will discuss students' competence development in the context of an interdisciplinary user innovation course carried out within higher engineering education at the University of Vaasa, Finland. In the chapter, we will firstly reflect intended course competence development toward earlier research findings regarding crucial engineering competencies, and discuss how pedagogical choices affect the development of competencies. Secondly, we present the course concept including course structure, content, and assessment of learning. Thirdly, we present the results of an analysis of the students' perceptions of their own learning of competencies and their best experiences from the user innovation course. Finally, in the conclusion we discuss some challenges related to the course concept and the potential of applying this approach to other elements of higher engineering education in order to prepare the students for meeting the demands of future engineering workplaces.

5.2 *Crucial engineering competencies and intended case course competence development*

There are several earlier studies on engineering competencies. Passow and Passow (2017) conducted an extensive systematic literature review in order to grasp the generic engineering competencies and their relative importance for professional engineering practice. For this chapter, we base the discussion about crucial engineering competencies on the findings of Passow and Passow (2017) since they include a range of earlier presented competence lists in their study. In the following, we define competencies along Passow (2008: 10) as "the knowledge, skills, abilities, and other characteristics that enable a person to perform skillfully (i.e., to make sound decisions and take effective action) in complex and uncertain situations such as professional work, civic engagement, and personal life."

Based on their extensive review on engineering competencies, Passow and Passow (2017) define the following 16 important generic engineering competencies: (1) solve problems, (2) communicate effectively, (3) coordinate efforts, (4) take initiative, (5) think creatively, (6) take responsibility, (7) measure accurately, (8) interpret data, (9) define constraints, (10) devise process, (11) gather information, (12) expand skills, (13) make decisions, (14) design solutions, (15) apply knowledge, and (16) apply skills. Passow and Passow (2017) discuss how this list of competencies relates to earlier presented ones, such as the competencies on ABET's list (ABET, 2014), and argue for the modifications and additions of them. In comparison with the list of ABET, Passow and Passow (2017) use *coordinate efforts, take responsibility,* and *devise process* instead of *teamwork, ethics,* and *manage projects.* They further split ABET's competence of *lifelong learning* into *gather information* and *expand skills.* ABET's competence of *design experiments* is expanded to *measure accurately,* which they separate from *interpret data.* Passow and Passow (2017) listed competence *define constraints* includes aspects of the competences *contemporary issues* and *impacts* from ABET's list. Finally, Passow and Passow (2017) add the crucial competencies *take initiative, think creatively,* and *make decisions.*

The case described in this chapter is an introduction to the phenomenon of user innovation. In the chapter, we will not go deeper into the phenomenon of user innovation as such, since the chapter will focus on fostering crucial engineering competencies. Regarding the phenomenon, we shortly state that the theoretical part of the case course is based on materials developed by the world leading user innovation scholar Eric von Hippel and his team at MIT, and that we use the definition and descriptions of the phenomenon developed and presented by Von Hippel. We thereby see a user innovation as an innovation developed "for personal or in-house use" (Von Hippel, 2017). Thus, user innovators can be free innovators as well as profit-seeking individuals and firms. In our course context, the user innovators were students developing innovations as a part of an interdisciplinary team, as will be further described in Section 5.3.

For fostering crucial engineering competencies, the user innovation course context plays a role in the sense that this kind of a course can be expected to foster certain competencies listed by Passow and Passow

(2017) more than others. These competencies can also partly be derived from the description of intended learning outcomes, which will be further discussed in Section 5.3. Innovation requires creativity by nature and the competence *think creatively* is therefore strongly supported by the course content. The course concept further supports the competence *coordinate efforts* since an essential part of the course is a team project. Besides *teamwork skills* and *innovativeness, initiativity* is mentioned in the description of intended learning outcomes as a working skill developed during the course. In Passow and Passow (2017), *initiativity* is expressed as the competence *take initiative*.

As defined above, user innovation is about users developing innovations for personal use, and the starting point for the innovations developed are typically problems the users face in their own lives. Based on this, it can be stated that the competence *solve problems* is also strongly linked to the course content. In the description of the intended learning outcomes, the aspect of cross-functional teams is further mentioned, i.e., the aspect of working interdisciplinary. *Interdisciplinary teamwork* is not mentioned as a competence of its own in the competence list defined by Passow and Passow (2017), it is included in the competence *communicate effectively* which is defined in the following way: "Communicate effectively with people that have diverse goals and backgrounds—across disciplines, organization levels, and organizational boundaries, through listening, oral, written, and graphical means" (Passow & Passow, 2017). In summary, the course content and context can be expected to develop the competencies *solve problems, think creatively, coordinate efforts, take initiative,* and *interdisciplinary teamwork* as a part of the competence *communicate effectively*.

For educators, an important question arises: How do pedagogical choices affect the learning of these crucial competencies? Alavi (1994) identified three characteristics of effective learning, namely (1) active learning and construction of knowledge, (2) cooperation and teamwork in learning, and (3) learning through problem-solving. These characteristics guide our case course. Further, contemporary pedagogical approaches underline that students should actively engage in their own education. They should learn in an environment that favors activity and experience, and foster immediate engagement (Biggs, 1996; Nordstrom & Korpelainen, 2011). Moreover, educators should use "learning tools that promote learning and include verbal, digital, visual or emotional tools which are used to increase personal and group commitment" (Nordstrom & Korpelainen, 2011).

Quite often the innovative, novel solutions to complex problems are not found within disciplines but in-between disciplines. The ability to integrate disciplinary knowledge is essential for today's complex problem-solving. This interdisciplinary thinking can be defined as

> The capacity to integrate knowledge and models of thinking in two or more disciplines or established areas of expertise to produce a cognitive advancement—such as explaining a phenomenon, solving a problem, or creating a product—in ways that would have been impossible or unlikely through single disciplinary means.
>
> *Boix Mansilla et al. (2000)*

For the students, interdisciplinary learning is interesting and improves students' understanding of theoretical concepts in a practical context (Gero, 2017). Interdisciplinary projects also support high academic engagement due to participation (Koch et al., 2017). Thus, working in interdisciplinary teams and learning interdisciplinary thinking are useful tools for developing engineering competencies. Interdisciplinary thinking most likely develops in a safe and supportive learning environment that is gained through an implementation of the constructive alignment principle (Spelt et al., 2015). The main idea behind this principle is to form a coherent configuration between assessment, teaching and learning activities and intended learning outcomes (Biggs, 1996). Moreover, in order to foster problem-solving and deep learning, Nordstrom and Korpelainen (2011) suggest a use of nonconventional tools such as drama, video, posters, model making, and other similar means.

Next, we turn into our case course—an interdisciplinary course on user innovation. We explain the constructive alignment of the course and describe the engaging learning activities that aim to promote deep learning and complex problem-solving in this specific context.

5.3 User innovation case course concept

The interdisciplinary user innovation course introduces the students to the phenomenon of user innovation and its basic concepts and tools, and this theoretical basis is applied in a team project. The courses at the University of Vaasa, Finland are in general given for students attending one and the same bachelor's or master's program, in this respect the user innovation course is different since it targets students from all disciplines ranging from engineering disciplines to business studies, communication studies, and administrative sciences. Most of the participants have no previous knowledge of user innovation and the theoretical part of the course, therefore, concentrates on the basics. The course was piloted in autumn 2016 and was given as a regular course for the first time in autumn 2017. At both times, there were around 60 students resulting in ten user innovation teams.

5.3.1 Intended learning outcomes

The course concept involves several intended learning outcomes focusing on user innovation. After completing the course, the student should be able to (1) identify, explain, and compare basic concepts of user-centered innovation, (2) explain the different parts of the user-centered innovation process, (3) apply the tools of user-centered innovation, (4) create a user-centered innovation concept in teams, and (5) evaluate the success of user-driven innovations. On top of these the course intended to develop particular working life skills, namely initiativity, teamwork skills, and innovativeness and creativity.

5.3.2 Course content and structure (teaching and learning activities)

The user innovation course is a 15-week course. The structure of the course is visualized in Figure 5.1. There are two face-to-face meetings with the whole class: one in the first week and one in the second last week. Between the face-to-face meetings with the whole class, two tracks support the students' learning process. The first track is formed by individual assignments based on videos and other materials available in the university's digital learning management system, Moodle. Via assignments, carried out in a traditional format of written assignments, the participants learn about the basic concepts of user innovation and the user innovation process. As the course is optional for the students, these individual assignments also make sure that the students are ready to invest their time for the course.

The second track consists of the user innovation team project carried out by interdisciplinary teams with 5–6 team members. The teams are formed in the first face-to-face kick-off, and the teaching team makes sure that all teams have members representing different disciplines. During the course, the teams are coming up with and developing a user

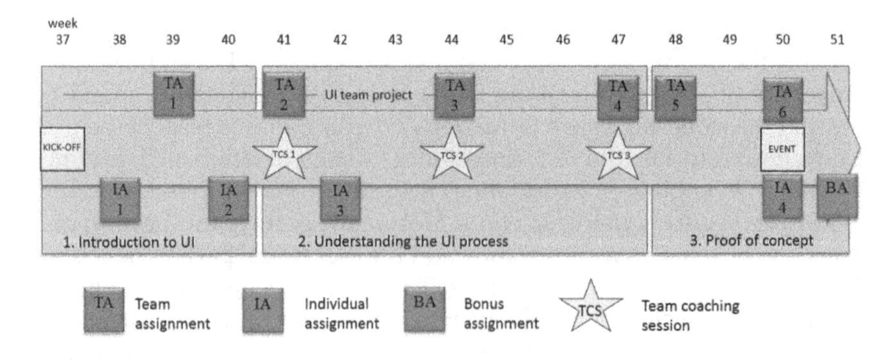

Figure 5.1 User innovation course structure.

innovation idea. The teams take responsibility for the process, but the development of the project is supported by three coaching sessions on the way. In these sessions, the teams present the current status of the projects based on a few questions for a panel of 3–5 interdisciplinary teaching team members. The sessions are 20–30 min and give the students an opportunity to develop their presentation and pitching skills as well as argumentation skills. The sessions further provide the teams with a possibility to address and discuss challenges of their teamwork. The final face-to-face meeting is organized as a "trade fair" where the teams present their ideas publicly. This fun event is a highlight of the course. In the last week, there is a collection of course feedback as a bonus assignment. The students individually reflect upon their learning process, the challenges faced during the process, and future entrepreneurial possibilities of the course teamwork.

The starting point in autumn 2016 was that the user innovation course would be conducted digitally. However, already in the development phase, the teaching team realized that this would be a challenge since the teamwork in general and interdisciplinary teamwork especially, is difficult to carry out without any face-to-face sessions and discussions (Jaskari & Jaskari, 2016). In the pilot, we carried out the meetings with the whole class and the teams as face-to-face sessions resulting in five face-to-face sessions during the course. In the pilot feedback, the students were in general very satisfied with the course, but they hoped for further possibilities to attend the course remotely. In autumn 2017, the coaching sessions were partly carried out remotely. We kept the kick-off as a face-to-face event since we see a tendency that students attending the kick-off face-to-face are more dedicated to attend the course. The final event has also an important function as a face-to-face session since it is the first opportunity for the teams to get to know the projects of the other teams. Figures 5.2 and 5.3 give

Figure 5.2 Students presenting their user innovation team projects in the final course event. (Courtesy of Minna-Maarit Jaskari.)

Figure 5.3 Students presenting their user innovation team projects in the final course event. (Courtesy of Minna-Maarit Jaskari.)

impressions from the final event in December 2017, where the students presented their concept-of-proof in a fun and exciting way.

5.3.3 Assessment of learning

The assessment of learning aimed at evaluating both the individual and team learning. Individual assignments were mainly the written assignments based on videos and other materials, additionally; the participants' presentations in the final event and their reflections upon the course in the final course feedback were individually assessed during the course. These assignments were constructed to assess the intended learning outcomes of (1) *identify, explain and compare basic concepts of user-centered innovation,* (2) *explain the different parts of the user-centered innovation process, and* (3) *evaluate the success of user-driven innovations.* The intended learning outcomes of (4) *apply the tools of user-centered innovation and* (5) *create a user-centered innovation concept in teams* were assessed through the team project in different steps. In the first team assignment, the teams presented themselves and their first project ideas in form of a video. In the following three team assignments, the teams presented the current stage of the project in a document and in an oral presentation for the teaching team in the coaching sessions. The last two team assignments were a poster presenting the teams' user innovation projects and a video proofing the concepts of the teams. A detailed description of the assessment of the assignments can be found in Appendix 1.

5.4 *Student perceptions on competence development and best experiences*

In order to get feedback from the students and to understand how students perceived the competence development, we included a post-course evaluation as a bonus assignment. In the evaluation, we asked some general questions such as what were the best three aspects of the course, how it was to work in interdisciplinary teams and how they would like to develop the course further. We also asked some specific questions concerning course management, such as online vs. offline working methods, assignments, and need for more precise information about the course. More specifically for this chapter, we wanted to analyze how well this course supports the development of competencies mentioned in the intended learning outcomes of the user innovation course and the crucial engineering skills listed by Passow and Passow (2017). In the following, we summarize the essential results of the course evaluation 2017 regarding student perceptions of competence development and student feedback regarding best experiences from the course.

5.4.1 *Competence development*

We were interested to see how the students perceive their competence development during the user innovation course and asked the students, as a part of the post-course evaluation, to rate, "how did the course support the development of following competencies/skills relevant for working life?" on a scale *Very much, Above average, Average, Below average,* or *Very little.* We based the list of competencies/skills on Passow and Passow (2017) with some smaller modifications. Since the course focused on teamwork and not on project management as such, we decided to exclude the competence *devise process* as it, in our opinion, did not apply to the course concept. On the contrary, we added *time management, interdisciplinary teamwork,* and *cross-cultural skills* in the list. All of these are included in the competencies listed by Passow and Passow (2017), but since we could not include long definitions in our list in the evaluation, we found it important to explicitly include these in the list. As described in Section 5.2, Passow and Passow (2017) competence *communicate effectively* includes both the aspect of interdisciplinarity and the cross-cultural aspect, and *time management* is seen as a part of the competence *define constraints.* During the course, cross-cultural skills turned out to be very crucial for successful teamwork. The course was carried out in English, about half of the participants were international students, and all teams were a mix of Finnish and international students.

Overall 42 students out of total 57 course participants in 2017 responded to the post-course evaluation, which was a bonus assignment

Table 5.1 Percent of respondents rating the specific competence/skill as very much or above average

Think creatively	95.2%
Design solutions	90.5%
Solve problems	73.8%
Interdisciplinary teamwork	73.8%
Cross-cultural skills	73.8%
Take initiative	73.8%
Coordinate team's efforts	71.4%
Apply knowledge	71.4%
Make decisions	69.0%
Communicate effectively	69.0%
Take responsibility	66.7%
Apply skills	61.9%
Time management	57.1%
Gather information	47.6%
Define constraints	45.2%
Expand skills	45.2%
Interpret data	35.7%
Measure accurately	28.6%

as described in Section 5.3. Table 5.1 shows the share of respondents answering the question "How did the course support the development of following competencies/skills relevant to working life?" with *Very much* or *Above Average.*

Table 5.1 shows that there is a clear top two in the students' evaluation: The competences *think creatively* and *design solutions* are ranked as *Very much* or *Above average* by more than 90% of the students. Not just the competence *think creatively,* which is a very natural one in a user innovation course context, but also the competence *design solutions* is in the students' opinion a competence which is strongly developed during the course. Next, there is a group of four with the same share, *solve problems, interdisciplinary teamwork, cross-cultural skills,* and *take initiative.* All the central competencies or part of competencies discussed in Section 5.2 as expected ones, based on course content and context, can be found in the first two groups, and we can see that the development of these competencies/skills is, in a clear student majority's opinion, developed at least above average. A group of further four competencies with a share of around 70% follows, *coordinate team's efforts, applies knowledge, make decisions, and communicate effectively.* The fact that we added *interdisciplinary teamwork* in our list might have affected the responses to the competence *coordinate team's efforts,* but it still gets a strong support in the students' evaluation.

Three more competencies, *take responsibility, apply skills,* and *time management,* are ranked as at least above average by 57%–67% of the students. The remaining five competencies *gather information, define constraints, expand skills, interpret data,* and *measure accurately* are by less than half of the respondents seen as developed at least above average during the course.

In summary, the user innovation course is, according to more than half of the respondents, developing ten of the competencies mentioned in Passow and Passow (2017) list at least above average, and *interdisciplinary teamwork, cross-cultural skills,* and *time management* can be added to this group. There are just five of the listed competencies, which less than 50% of the respondents rank as at least above average. Based on the students' perception on competence development expressed in the course evaluation, the course concept and structure seem to support the development of crucial engineering competencies very well.

5.4.2 *Best experiences*

Regarding other student feedback on course contents, working methods, etc., our evaluation clearly shows that the students welcome this kind of course where the application of knowledge and skills is in focus. Students like creative and unconventional modes of learning, and the project course of user innovation that we taught provided this to the students. The overall evaluation is very positive: "The course is a great opportunity where students are pushed to think out of their comfort zone." When we analyzed the students' comments regarding the three best aspects of the course, the following categories can be formed:

The topic itself. The students brought up that the topic of user innovation itself was interesting and worth learning. The course incorporated MIT Open course material using video lectures by "the father of user innovation," Professor Eric von Hippel, and this aspect was taken up several times. The students also mentioned that the theoretical aspect was very interesting as students are exposed to real-life scenario while reading.

The students valued the pedagogy that put the theoretical **knowledge into practice**, whether it concerned the practical application of knowledge, real-life innovation, or analyzing user needs and creating solutions for users. Some of the students brought up how they thought that this kind of active learning is the best for them in order to thoroughly understand and deepen the learning. Also, the creative input required to accomplish the goal was appreciated as one student explains: "I really like to be creative and go wild with the ideas and that is exactly what we could do during this course!" (B–E)

One of the core ideas of the course is to give students an experience of working with different kinds of people. This was noted as one of the best aspects of the course. **Interdisciplinary** teams, versatile groups, spirit were aspects mentioned. For example, "That I got my team chosen for me so I meet five new people from different environments and we were able to confront our different way of thinking," or: "Fellow students in the group were from different faculties. In my opinion, this kind of group work was a wonderful opportunity to interact, exchange ideas and learn." (G–K)

The course structure along with different learning activities was mentioned several times, for example, the structure and timeline of the course, video lectures, no exam, and interesting individual assignments. Also, the balance between individual and group assignments were considered good. The students appreciated the flexibility of the course meaning working both online and offline and having only some face-to-face meetings. As the students came from three different faculties and from both bachelor and master's level, there are some difficulties in organizing team meetings. Thus, the flexibility allows students to coordinate their own work. Two aspects of the course management were mentioned several times as best aspects of the course. The first was the **coaching sessions**, where the teams presented their ongoing work to faculty members. The second was **the final event**, the trade-fair like presentations of teams' creative solutions. "The final event where we presented our concept was really inspiring, at least to Finns like me it is good to get used to marketing your own ideas" (L–O). The final event seemed to give the students a sense of accomplishment and pride.

5.5 Discussion and future perspectives

This chapter has presented a case study of a user innovation course aiming to develop crucial engineering competencies. Even though there is always need for continuous development, the students perceived this pedagogic structure to support the development of these competencies.

The course was able to engage students both emotionally and cognitively. The course was described to be fun and motivating but cognitively challenging at the same time. Indeed, at its best, this is exactly what active learning methods can support. The development of interdisciplinary thinking skills was evident as the best teams were able to use interdisciplinary thinking in their development process—the outcome was such that knowledge and skills from different disciplines were used to solve the problem. However, not all teams were able to reach this level. In these cases, the outcome of the course was dominated by one discipline. This is not a surprise—interdisciplinary thinking is not an easy task and rarely taught during one course. However, this course may have been a good starting point for many.

The students that were not satisfied with the course usually faced problems with the teamwork. The challenges within teamwork could rise from different sources. Some teams faced problems with unmotivated team members, some teams had problems with cultural or disciplinary differences, and others had team members that did not have time to complete the course. As the course was optional, it was easy to drop out without any consequences. The teams were encouraged to bring up the problems as early as possible, but in many cases, this did not happen.

The advancement of interdisciplinary thinking can be challenging. Discussions among team members require, on the one hand, presenting own ideas and arguing for them, on the other hand, being willing to listen to others' ideas and compromise, when necessary. Based on our experience, we emphasize the need for some face-to-face events, such as coaching sessions, where these challenges can be discussed openly. When the teachers, for example, explicitly ask about how the teams have resolved typical challenges faced in interdisciplinary teamwork, the students understand that it is not only their team that has problems and that there are ways to overcome these challenges.

We as engineering educators are responsible for providing such challenges for our students so that the crucial engineering competencies are developed already during education. In this chapter, we have provided an example of a user innovation course, where students perceived the development of some of these crucial competencies. We hope that this example would inspire other engineering educators to use and develop this concept further.

Appendix 1: Description of course assignments and their assessment in 2017

		Deadline	Marks
IA 1	Individual assignment 1 written assignment	22.9.2017, 16.00	0–10
TA 1	Team assignment 1 team presentation video creation of common document	29.9.2017, 16.00	accepted/not accepted
IA 2	Individual assignment 2 written assignment	6.10.2017, 16.00	0–15
TA 2	Team assignment 2 ppt presentation from TCS 1 team meeting notes as pdf	13.10.2017, 16.00	0–10

(*Continued*)

		Deadline	Marks
IA 3	Individual assignment 3 written assignment	20.10.2017, 16.00	0–15
TA 3	Team assignment 3 ppt presentation from TCS 2 team meeting notes as pdf	3.11.2017, 16.00	0–10
TA 4	Team assignment 4 ppt presentation from TCS 3 team meeting notes as pdf	24.11.2017, 16.00	0–10
TA 5	Team assignment 5 UI project poster	1.12.2017, 16.00	0–5
IA 4	Individual assignment 4 attendance face-to-face event	13.12.2017	0–10
TA 6	Team assignment 6 UI project presentation video	15.12.2017, 16.00	0–15
Total			0–100

References

ABET (2014). Criteria for Accrediting Engineering Programs, 2015–2016. Available at http://www.abet.org/wp-content/uploads/2015/05/E001-15-16-EAC-Criteria-03-10-15.pdf, accessed 15 April 2018.

Alavi, M. (1994). Computer-mediated collaborative learning: An empirical evaluation. *MIS Quarterly*, 18(2): 159–174.

Biggs, J. (1996) Enhancing teaching through constructive alignment. *Higher Education*, 32: 347–364.

Boix Mansilla, V., Miller, W.C., & Gardner, H. (2000). On disciplinary lenses and interdisciplinary work. In *Interdisciplinary Curriculum: Challenges of Implementation*, edited by S. Wineburg and P. Grossmann, 17–38. New York: Teachers College Press.

Boyatzis, R.E. (1982). *The Competent Manager: A Model for Effective Performance*. New York: Wiley.

Boyatzis, R.E. (2008). Competencies in the 21th century. *Journal of Management Development*, 27(1): 5–12.

Gero, A. (2017). Students' attitudes towards interdisciplinary education: A course on interdisciplinary aspects of science and engineering education. *European Journal of Engineering Education*, 42(3): 260–270. doi:10.1080/03043797.2016.1158789.

Henri, M., Johnson, M. D., & Nepal, B. (2017). A review of competency-based learning: Tools, assessments, and recommendations. *Journal of Engineering Education*, 106: 607–638. doi:10.1002/jee.20180.

Jaskari, H., & Jaskari, M.-M. (2016). Success factors in teaching strategic sales management–evidence from client-based classroom and web-based formats. *Marketing Education Review*, 26(3): 171–185.

Koch, F.D., Dirsch-Weigand, A., Awolin, M., Pinkelman, R.J. & Hampe, M.J. (2017). Motivating first-year university students by interdisciplinary study projects. *European Journal of Engineering Education*, 42(1): 17–31. doi:10.1080/03043797.2016.1193126.

Nordstrom, K., & Korpelainen, P. (2011). Creativity and inspiration for problem solving in engineering education. *Teaching in Higher Education*, 16(4): 439–450.

Passow, H. J. (2008). What competencies should undergraduate engineerings emphasize? A dilemma of curricular design that practitioners' opinions can inform, *Doctoral Dissertation*. University of Michigan. Available at https://deepblue.lib.umich.edu/bitstream/handle/2027.42/60691/hpassow_1.pdf?sequence=1&isAllowed=y, accessed 21 March 2018.

Passow, H. J., & Passow, C. H. (2017). What competencies should undergraduate engineering programs emphasize? A systematic review. *Journal of Engineering Education*, 106(3): 475–526.

Spelt, E. J. H., Luning, P. A., van Boekel, M. A. J. S., & Mulder, M. (2015) Constructively aligned teaching and learning in higher education in engineering: What do students perceive as contributing to the learning of interdisciplinary thinking? *European Journal of Engineering Education*, 40(5): 459–475.

Spencer, L.M., & Spencer, S. M. (1993). *Competence at Work: Models for Superior Performance*. New York: Wiley.

Von Hippel, E. (2017). *Free Innovation*. Cambridge, MA: MIT Press.

chapter six

Case study

Effective use of technology for classroom instruction—hybrid and online learning

Kailash M. Bafna
Western Michigan University

Contents

6.1 Introduction: The need for change

The method of delivering education has changed very significantly since the start of this millennium. This change has been precipitated partly by the availability of new technology and applications software, and partly to satisfy the needs of the generation of students who are now in college—better known as the millennial and centennial students. These students have some unique characteristics which make it difficult for them to derive maximum benefits from the traditional classroom lectures of 50–75 min duration. Research suggests that millennials and centennials prefer a variety of active learning methods. When they are not interested in something, their attention quickly shifts elsewhere. Interestingly, many of the components of their ideal learning environment—less lecture, use of multimedia, collaborating with peers—are some of the same techniques research has shown to be effective.[1] This indicates that the typical "chapter" format of lectures should be modified into smaller "learning units," each unit being a small topic related to the overall chapter and having specific learning outcomes.

These students have grown up by being able to Google anything they want to know; therefore; they do not typically value information for information's sake. As a result, the professor's role is shifting from disseminating information to helping students apply the information. One of the greatest challenges for teachers is to connect course content to current culture and make learning outcomes and activities relevant.[1] This indicates the need for helping students apply the information learned in the course so that students can understand the value of that information. In the traditional education system where the information is delivered to the students in the classroom through lectures, there is not enough classroom time available to help the students with the application of the material. This need has resulted in some educators "inverting" the classroom. Lage et al.[2] state that

> Inverting the classroom means that events that have traditionally taken place inside the classroom now take place outside the classroom and vice versa. The use of learning technologies, particularly multimedia, provides new opportunities for students to learn, opportunities that are not possible with other media. . . . The instructors focus on the desired outcome (for instance, having the student prepared for discussion) and allow the student to choose the best method to reach that outcome.

Although the flipped classroom can be taught without online resources, the Internet allows the instructor to invert the classroom without sacrificing content coverage.[3]

Take the example of Khan Academy. Originally developed as a means of helping his cousins with math, Salman Khan's efforts have expanded into 1,800+ videos on YouTube, with nearly 22 million views between them. In these brief 10–15-min tutorials, Khan explains basic (and not so basic) math concepts in a concise manner that students can easily digest and reference later. Khan Academy videos are viewed more than 70,000 times per day—that's more students than most major universities.[4] The popularity of Khan Academy has even drawn raves from Bill Gates, among others—suggests that mini-lectures, delivered apart from the classroom, could pick up momentum in higher education.[5] The success of the Khan Academy is an excellent example of the acceptance of creating short educational videos of individual concepts, and posting them on the web for students to view and learn.

6.2 Needs of the millennial learner

The millennial generation (or generation Y) includes those born between the years 1980 and 2000, and the centennial generation (or generation Z) includes those born between 2000 and the present. Since the beginning of this millennium, we have been educating the millennial generation in our colleges and universities and, starting this year we will also start educating the centennial generation for almost another two decades. Both these generations have grown up around technology and enjoy brief and extended encounters with computers, MP3 players, iPods, iPads, tablets, and smart phones.

As shown in Figure 6.1, today's students have developed several unique characteristics because of growing up in this digital age. The distractions provided by the large variety of technology-related gadgets and social media have resulted in these individuals developing shorter attention spans, thus acquiring multi-tasking capabilities. These generations have grown up with their parents keeping them busy by involving them

Figure 6.1 Characteristics of today's students.

in a lot of after-school activities. This has resulted in the parents having to juggle the time management for their children, leaving them with poor time-management skills in college. Also, having participated in numerous group extra-curricular activities, and due to their affinity for social media, they enjoy working in teams with their peers. Technology has taken over handwriting, and short messaging is now preferred to more formal writing. Since the cost of education has skyrocketed in the past three decades, a larger percentage of today's students are working part-time to pay their way through college. The quest for knowledge for this generation of students is instantly satisfied through Google searches and they are quickly moving away from textbooks as a source of information. Due to the availability of a variety of social media, they have developed a need for instant feedback and gratification. One can easily notice that every waking minute in the lives of these millennials and centennials revolves around digital technology. As such, they are increasingly being referred to as "digital natives" or "Google kids."

6.3 The changing methods of education

Although we have used the traditional methods of classroom education for several centuries, because of the availability of technology, several new methods have been introduced in the past 30–40 years. As shown in Figure 6.2, the three most commonly used newer methods are online education, flipped classroom education, and hybrid education. *Traditional Education* is held in a classroom and the instructor usually faces the students and delivers the lecture (body of knowledge). The students apply this body of knowledge through homework assignments, by working on projects, solving problems, and writing reports outside the classroom. In *Online Education,* the individual students learn at their own choice of location (usually their place of residence) at a time convenient to them. They are given a lesson plan by the instructor at the beginning of the semester/quarter along with other reading materials (which may also include a text book) and they follow it as best as they can depending upon their time availability and other commitments. Tests may be given periodically, or submission of some reports may be required. In the *Flipped Classroom Education* model, the instructors record their lectures and post them on an institution's server. These videos can then be accessed by students on the web. The instructor may release them according to a given schedule or all at one time. The students view these "lecture videos" individually according to a given schedule but at a time convenient to them. If necessary, they can view any part of the video multiple times until they understand the material clearly. The students then meet the instructor in the classroom on a regularly scheduled basis. During these class meetings they can clarify any materials on the "lecture videos," and apply the

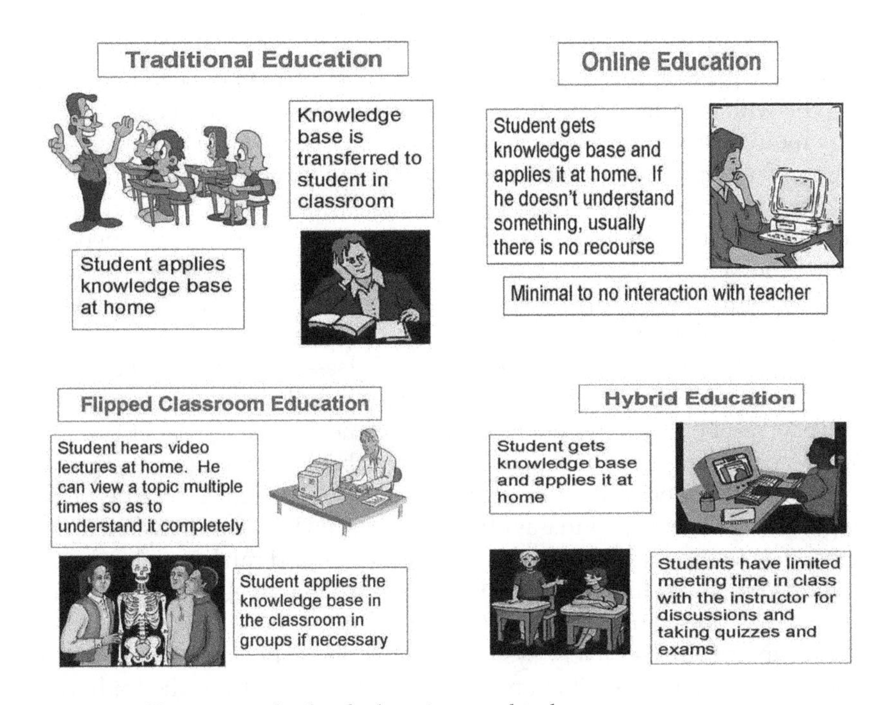

Figure 6.2 Various methods of education used today.

learned materials to work on projects and solve problems individually or in groups. In the *Hybrid Education* model, students acquire the knowledge base at home by viewing the "lecture videos" but also apply the knowledge base by working on projects and solving problems at home. To help the students in the application of the knowledge base, the instructor may post additional videos online showing the solution of sample problems or application of the learned materials to other areas. The instructor then meets the students in the classroom on a limited basis to answer any questions from the students and for testing purposes.

My innovative method of teaching uses a combination of all the three newer techniques described above. I have found this technique to be extremely beneficial for educating the millennials. Moreover, I expect that following this technique will help the student in their lifelong learning experience.

6.4 *The Engineering Economy course*

This case study describes my experiences in teaching the subject of Engineering Economics, a mathematically based course. For almost 40 years, I have been teaching one and up to three courses covering the subject of Engineering Economics at the undergraduate level for (1) all engineering majors, (2) all mechanical engineering students, and (3) for technology majors. In these courses, I teach several different mathematical concepts dealing with the time-value of money. These concepts are then used to compare multiple alternatives to select the best one. This course also covers the concepts of depreciation and its computation by several different methods. Additional topics such as impact of taxes on earnings (personal and corporate), the effects of inflation, and risk analysis are also covered depending on the course. It is primarily a mathematical course. I started incorporating various features of the university's learning management system (generally referred to as eLearning) into these courses 14 years ago. For the past 7 years I have offered both these courses as hybrid classes and, on two occasions, as online classes. Since 2011, both courses are handled entirely through the university's eLearning system and have become completely paperless. As a result, not a single sheet of paper is exchanged between the students and me.

The remainder of this chapter describes the current structure of the Engineering Economy course as offered by me, and how it is delivered to students. The techniques described are simple to use and can be easily adapted to most courses. All it requires is the availability of an eLearning platform in the educational institution, and a desire by the individual instructor to change their course delivery system. The chapter concludes with a discussion of the effectiveness of changing the methodology from the traditional classroom to the hybrid format and the benefits observed by the students.

6.5 Restructuring the course for flipped classroom delivery

Although textbooks are typically divided into chapters based on major topical content, I have divided the course material into much smaller units based on learning outcomes from the course. Hence, the entire course has been divided into about 40 learning units, each unit typically being a single topic. For each learning unit, I have prepared four items: (1) PowerPoint notes, (2) "lecture videos" based on my notes, (3) some solved examples which are also presented in video format, and (4) several assigned end-of-chapter problems along with answers for students to solve. Some of the introductory units only have the first two items mentioned above. Students are advised to complete all the parts of a single learning unit before they move on to the next unit. They have found this "structure" to be valuable since they can now completely familiarize themselves with one unit before moving on to the next. Each of the above four components of a learning unit are described below:

PowerPoint Notes: I have spent a lot of time in preparing the PowerPoint slides for each learning unit and there is nothing unique about this. However, since this is a mathematical course, I have taken the effort of introducing animation and colors into each slide. The animation during the lecture allows equations to be viewed one part at a time. Thus, the students can participate in the formulation of the equation as it is being presented. Moreover, I have used different colors to depict the various cash flows in the graphics of the cash flow diagram. Each part of the equation has also been coded in the same color so that students can correlate a cash flow in the diagram with its corresponding part in the equation. Students have commented that this makes it easier for them to understand the material. All my PowerPoint notes are posted on eLearning as PDF files, so students can download them and use them to follow the lecture videos.

Lecture Videos: With the use of a screen capture application (initially I used *My Screen Recorder Pro* but, since the last 4 years, I have switched over to *Camtasia Studio* because it allows me better video editing capabilities), I have developed the ability to record my "lecture videos" directly on my laptop without any assistance. I can record lectures in my office or at home and the only equipment required is a small microphone and the laptop computer. I have even found the microphone built into my laptop to be adequate. The applications mentioned above record all screen movements with my voice as I move through my PowerPoint presentation. The recording (video and audio) is converted to an MP4 file and is loaded on the university's server for streaming. The students can easily view the video on any computer and can fast forward and rewind the video without any significant delay. This allows them to view any segment of the video multiple times, if necessary, to fully understand the material.

I am currently using my fourth generation of videos. Initially, these videos were constrained by the duration of the class of 50 min in length and could only be viewed on a computer. Based upon student feedback, I have now shortened the duration typically to about 15 min so that a student can easily complete the viewing in one sitting. Besides the computer, the student can now view the videos on a tablet or even a smart phone. I had to modify my PowerPoint slides with less material on each slide so that it could be viewed on smart phones also. If any topic is longer, I divide it up into two or three units so that the "video lecture" remains within the length guidelines.

Videos of Solved Examples: For each learning unit, I have solved 2–4 problems and prepared a video of the solutions. Before creating the video, the questions are entered into the computer. Below each question, all the given data are summarized followed by a solution to the problem. When recording the video lecture, I first read out the question, and then analyze the information that is given followed by a discussion of the analysis leading to the solution methodology. Once again, this video is loaded on the University's server and is streamed on demand. Just like the lectures, the student can view any part of it multiple times by rewinding or fast forwarding it as desired.

Assigned Problems: For each unit, I assign a list of 4–8 problems from the end of the chapter in the text book related specifically to that unit. I also give the answers for each of the assigned questions. The student is asked to solve all the assigned problems before considering the unit to be complete. No submissions are required and hence there is no grading involved. I have found that very few students solve all the assigned problems, but the more problems they can solve, the better it is for them.

6.6 Introducing the use of technology for easier problem-solving

Students typically solve engineering economy problems by using either formulas or interest tables, values of which are usually given at the back of the textbook. Alternatively, the students buy a financial calculator just for this course (since engineering students use scientific calculators for their other courses). Although traditionally the primary focus in Engineering Economy textbooks has been to solve problems using formulas and interest tables with only a brief reference to solving them using Excel, I have developed the entire course where all problems are solved using Excel. However, as Excel is only a problem-solving tool and does not teach the concepts of time-value of money, I have developed a unique combination of solving problems with formulas, interest tables, and Excel in the first one-third of the course to allow the students to grasp the concepts of time-value of money. I then transition entirely into solving problems using Excel only in the remaining two-thirds of the course. Students need their text books (for the interest tables) and calculators for the first part of the course but do not need them in the second part of the course.

Many engineering students in the United States take the national Fundamentals of Engineering (FE) exam (as a first step to becoming a Registered Professional Engineer), which has several questions on engineering economy. Since students are only allowed to use formulas, interest tables, and calculators during this exam, teaching the first part of the course by the traditional method for solving problems gives the students enough background to take the FE exam. However, due to the major change of using Excel to solve all the problems in the course, its contents are much more meaningful to the students for their future as they can use Excel even after graduating (since text books and calculators are no longer required to solve the problems).

6.7 Using eLearning for delivering the course

Until 2011 the eLearning system, *Blackboard Learning System Vista Enterprise (WebCT Vista)*, was used at my university and was provided by *Blackboard Inc.* At that time, the university switched over to a different system provided by a Canadian Company, *Desire2Learn*. As mentioned above, I started using some aspects of eLearning 13 years ago and, currently, I am using several tools to make the course materials available to my students and for administering examinations and quizzes. These tools from eLearning allow me to offer my course in the flipped classroom, hybrid, and online formats. The purpose of each of these tools and how they are being used in my course is described below.

Contents: All the material needed for the course is placed under contents. This material can be in the form of files or can be linked directly to files. My PowerPoint notes and the list of assigned problems with answers are loaded as PDF files. All the lecture videos and solved problem videos are loaded on the university's streaming server and their URLs are linked to each topic under contents. As such, when this link is clicked, the video automatically opens in a box ready for playing. Each file can either be made available for the entire semester or it can be made available during a window between any two dates. I make all my course materials (except quizzes and exams) available for the entire semester. The course syllabus is also placed as a PDF file under contents. A sample course schedule is shown in Figure 6.3.

As shown above, the student is given a weekly schedule, i.e., all the items that should be completed during each week of the semester with a reference to the modules and the chapters in the textbook. Since the typical student has difficulty in following a schedule (because they spend their time on what is interesting instead of what is required), I have instituted a weekly quiz which the student is required to take once they have viewed the video lectures for that week. To give the students more flexibility in

Week No.	Week Of:	Module (Chapter)	Self-Evaluation Quizzes[1] (Last Date)	Quizzes[2] (Date)	Exams[3] (Date)	Assignments[4] Dates
1	1-9	0, 1, 2				
2	1-16	3 (3)	SEQ1 (1-22)			
3	1-23	4 (4)	SEQ2 (1-29)	TQ (1-25)		
4	1-30	4 (4)	SEQ3 (2-5)	Q1 (2-1)		
5	2-6	5 (5-9)		Q2 (2-8)		A1 Post 2-6 A1 due 2-9
6	2-13	6 (5-9)	SEQ4 (2-19)		E1 (2-15)	
7	2-20	7 (5-9)				
8	2-27	8 (5-9)	SEQ5 (3-5)			
	3-6	Spring Break				
9	3-13	9, 11 (5, 9)	SEQ6 (3-19)	Q3 (3-15)		
10	3-20	12 (9)	SEQ7 (3-26)		E2 (3-22)	
11	3-27	13 (11)				A2 Post 3-27 A2 due 3-30
12	4-3	13 (11)	SEQ8 (4-9)			
13	4-10	14 (12)	SEQ9 (4-16)			
14	4-17	14 (12)	SEQ10(4-23)			
15	4-26	Comprehensive Final Exam (E3) on Wednesday, April 26, 8:00 AM – 9:40 AM				

Figure 6.3 Course schedule for the spring 2017 semester.

their schedule, they can take the quiz anytime during that week. Since these quizzes have a grade component to it, it helps to keep most students on schedule.

These are termed self-evaluation quizzes (SEQ) and more details about these SEQs are given in the next section. There are three quizzes (Q) and three exams (E) administered during the semester. These are described in more detail under Dropbox. The course includes two assignments (A) which are made available through the Dropbox and the students have 3–4 days to complete these assignments and submit their completed files to the Dropbox.

Quizzes: This module is used to administer tests of any kind. It can be made available either on a specified date and time or during a prespecified window. The testing material questions can be in a variety of forms such as multiple-choice, true, and false, etc., and can be computer graded with the grades being posted automatically to the grade book. I use this feature to administer weekly multiple-choice quizzes (SEQs) based on the learning units assigned for that week. Each SEQ can be taken by the student at any time during the week once they have reviewed the posted materials for that week. The test score is released to the student immediately upon completion of the quiz. I am now allowing two attempts for each SEQ (the same questions are administered in a random order for each attempt) and the higher of the two scores is posted for that SEQ. This encourages students to review the materials again before the second attempt. Each SEQ (10 in all) is worth 2% of the course grade.

Dropbox: If there is any evaluation material that needs to be hand-graded, it can be placed in the Dropbox. Each item to be submitted is placed as a unique item in the Dropbox and is released to the students on a prespecified date and time. The completed file must be submitted by the student at a predetermined time on the same date or any other date as specified. As shown in Figure 6.4, the testing/instruction file(s) to be released are posted in the Dropbox and, upon completion of the item, the students submit it back to the Dropbox. I use this feature for three different types of evaluations. During the semester, I give three one-question quizzes of 15 min each (Q1, Q2, and Q3) and three four-question exams of 50 min each (E1, E2, and E3). Since my class meets in a room equipped with computer workstations, these quizzes and exams are taken during the class meeting. Upon completion, I download all the student files (as a zip file) to my laptop, grade them on the laptop, and return the graded file back to the student with my comments. I also post the student's grade for that item and the solution to the quiz/exam being graded. This helps the student understand how it should have been solved. I also use the Dropbox feature for two take home assignments (A1 and A2). I place the assignment and a video of my instructions on how to work on the assignment (formatting and procedure). Once again, upon completing the assignment, the student submits their work back to the Dropbox.

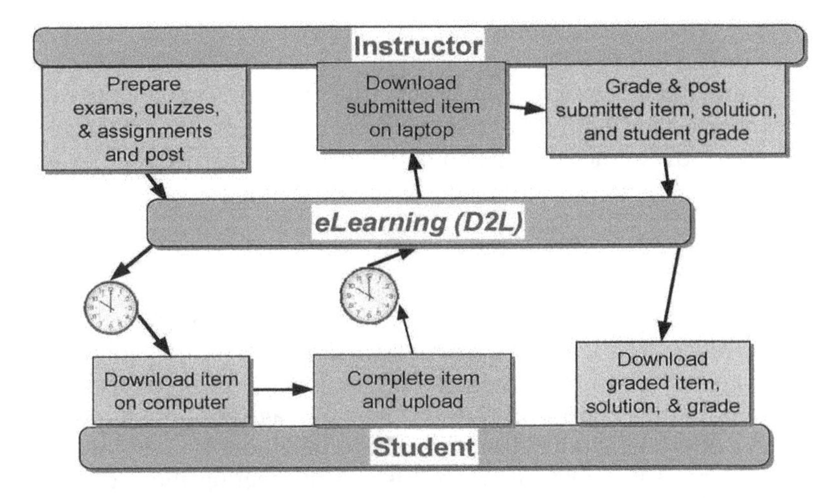

Figure 6.4 Administering and grading quizzes and exams using dropbox.

Calendar: This tool is used to compile a calendar for the course which indicates all the deadlines that have been established for the various items in the course. Since one of the hardest parts of self-study is to maintain self-discipline and have good time-management skills, a calendar with all the due dates shown is found by the students to be extremely valuable. My students have found the calendar tool to be very beneficial as they have only one location to refer to periodically to keep track of the course requirements.

Reports: At any point in time, the instructor can look at this tool to know how many students have visited each component of the unit and how much time has been spent on it on an average basis. Similar information is also available on any specific student. This helps the instructor to see if a student is keeping up with the course materials or is falling behind. I regularly monitor the reports tool and send emails to students who I feel are behind. This "forces" them to keep up with the material since they know that there is a "pair of eyes" watching them.

Grades: This is a very valuable tool for the students—the better students always want to know their grade with reference to the class, so they can try to improve, whereas the weaker students always want to know their grade standing as the semester progresses to see if they will pass the class or obtain a certain minimum grade. As I finish grading each item and post the grade on eLearning, the student knows their exact grade standing from the grade book. Since I have the option for the system to compute the grade based only upon the completed items, it gives feedback to the student about their current grade standing. The students really love this immediate feedback on a real-time basis and many of them take corrective action to avoid a disastrous semester end.

Email: This feature stores each student's email address so that the instructor can send emails to one or more students or even to the entire class to make them aware of any upcoming requirements for the course, or for the need for them to take some corrective action. I have found that using this tool is much faster and simpler than using my regular email system.

6.8 Using the flipped classroom model

During the last 7 years, I have offered 30 sections of Engineering Economy (there were three different courses designed for different majors in engineering and in technology) which were taught by me as hybrid classes and two of them as online classes. With an average enrollment of 25 students per offering, about 800 students have taken my classes. Two of the courses were of three credits each. Instead of meeting these classes for 3 h a week, I met the class once a week for 1 h only. As mentioned earlier, I have placed all my lectures on eLearning in the form of video lectures which the students view at their convenience. I do not lecture during the class period and this time is used exclusively for answering questions, taking quizzes and exams, and for solving additional problems. Earlier, when I used the traditional education model, my class time was primarily spent in lecturing on topics and, occasionally, solving a few problems. With this new format of video lectures, I have flipped the classroom. I have also created videos while solving 70 end-of-chapter problems spread out through all the topics and placed them on eLearning. Additional problems are solved in class when time permits. The students have found this format to be extremely valuable and are able to perform much better in their exams and quizzes, thus raising the overall class average.

For the two offerings of the course as an online class, I posted the same materials that I use for the hybrid class offerings. The only real difference between the online and hybrid offering is that I do not meet the online students at all throughout the semester. They are geographically dispersed due to their work locations. As such, they take exams and quizzes at the work place or at home. All the students in the class take these assessments at the same time, typically at 9:00 pm (after the children may have gone to bed and the enrolled students have some quiet time). Another difference is that if any online student has a question, they can communicate with me by email and, if necessary, I will call them on the phone to help them resolve their problem.

In its current form, all my notes and class materials are posted on eLearning. The student can look at the course materials and view the video lectures and solved problems any time at their convenience, thus offering them more flexibility. If they want to review their graded work or its solution, they can access it on eLearning. The class statistics for each quiz/exam are also available to the student so that they can compare their performance with that of the class.

Since I started using the flipped classroom model and various tools of eLearning, the course has become completely paperless. Simply speaking, although the students submit 18 quizzes, exams, and assignments during the semester, not a single sheet of paper is used in the solving, submitting, and grading process—hence, it is truly a paperless class. I have made it a point to grade every single submission and return it to the student with their grade through eLearning within 24 h. In one semester, students listed the following advantages of this paperless class:

- More convenient to go through notes. Teacher makes everything very organized and easy to study.
- It makes students use new skills and it forces us to adapt to change.
- Grades, notes, and resources are all stored in one place and are easily accessible.
- The PowerPoint notes, solutions, extra examples, and Excel are all very valuable. Don't have papers to keep organized. Everything is always available in one spot—online.
- I have access to course materials anywhere on campus, and at home. I like receiving instructions online and having assignments graded and posted online.
- I don't have to carry a binder around and I can just put everything on my flash drive.

Recently, in one semester I offered two sections of the same course—one section was offered as a hybrid class and the second section was offered as an online class. Since both sections were for on-campus students, the online students met with me in a classroom on six occasions during the semester to take the three quizzes (15 min each) and the three exams (50 min each). No problems were solved during these meetings. When I compared the results of both the sections at the end of the semester, I did not observe any significant difference in the results.

6.9 My journey from traditional to hybrid education

I started my teaching career in 1971 and, until the spring 2004 semester, I had taught all my courses in the traditional classroom format, meeting with the students three times each week for each three-credit course. Because of the large number of students who failed my classes (12%–25%), I was always concerned and would comment to my peers that the high schools are not doing an adequate job in preparing the students for college. I did not realize that I was dealing with a vastly different generation of students that had grown up in the digital age and needed innovative methods using technology in their learning.

Starting with the fall 2004 semester, I decided to use some form of technology (such as Excel) in my engineering economics course. As such, I started having all my classes scheduled in a computer-equipped classroom. Since each student had a workstation available, I began different types of experiments each semester. I first started using eLearning to release and collect the quizzes and exams in my Engineering Economy classes. This was a very major change for the students since nearly all of them were only used to taking exams on paper. It also resulted in my making a major adjustment in how I graded the exams. When they were held on paper, I would grade one question at a time for the entire class before moving on to the next question. When the students started taking the exams on the computer and submitted their completed files to eLearning, I had to open each students file to grade a question. It was naturally inconvenient to open each student's file multiple times to grade their exams (one question at a time) in the way I was used to. It took me a little getting used to but, very quickly, I adjusted to opening a student's file, grading all the questions in it, and then moving on to the next student's file, and so on.

In Engineering Economics, the questions were typically solved using formulas and factors, values of which had to be taken from interest tables at the end of the text book. The exam file consisted of multiple questions with one question to a page as is typically done when exams are prepared on paper. I started by preparing a file created in Word with as many pages as there were questions in the exam. When I wanted to experiment if students could solve a given question in Excel, I would create the question at the top of a page in Word and then I had to embed a blank Excel worksheet under the question on the same page. Initially this presented a problem for the students, but they soon got used to it.

When I started giving the exam to be completed on the computer, I had to find ways of preventing plagiarism. When the student was on the computer, they could open multiple files, including my notes and solved examples, email, etc. I had to find ways to prevent a student from emailing the exam (as an attachment) to a friend at some other location, have that friend solve the exam, and then return the file to the student. They could then just submit that file as their own work. To avoid this, I came up with the novel idea where on all the pages of the Word file, the entire page had a very light-colored background (with the color varying from exam to exam). During an exam, I would stand at the back of the classroom so that I could see the screens of all the students in the room. If I found that a student's screen changed colors, I would immediately walk up to them and find out what they were doing. I even informed the students at the start of the semester as to why I use a colored background on the exam pages.

Because of this technique, I found that the students became mindful and would not do anything they were not supposed to.

In subsequent years, I started teaching the entire Engineering Economics course using Excel for solving the problems. The students now did not need any formulas and interest tables (since these were already embedded as financial functions in Excel) and so the exams automatically became closed book. In a single Excel file, I could have multiple questions, one question on each worksheet. If the students needed some data (such as tax tables) to solve a question, I would provide them with this information on a separate worksheet titled "Data." I would enter the question in a text box at the top right of each worksheet and enter a colored background within the textbox. By standing at the back of the classroom, I could ensure that the students were working only on my exam (and no other files were opened).

I have been giving all my exams in Engineering Economics on the computer since 2008. As technology changed frequently, I also had to keep myself current to prevent students from taking advantage of me. I still remember a specific incident from a few years ago. A student approached me and asked if he could use Excel in Google Docs to save the exam file and then work on this file to complete the exam. After investigating, I found that someone else could collaborate and work simultaneously on the same file on Google Docs to solve the questions for the student in my class. My student would then submit the file as his own work. Obviously, I had to deny the request.

Whenever somebody wants to work at the cutting edge of technology, they must be prepared to face numerous problems, and must find solutions to these as they move along. I have only related some of my experience in this section. Along the way, I had several other experiences which are too numerous to elaborate here. However, as I was confronted with a challenge, I took it as an opportunity, did some self-education, and found a workable solution. I have found that my experiences in moving to the flipped classroom format and the hybrid class format of teaching have been very self-satisfying and have kept me energized.

6.10 Documented effectiveness of changing the methodology

In Figure 6.5, the distribution of individual grades in the course (A, BA, B, CB, C, DC, D, and E) expressed as a percentage of the total class size in selected semesters in my Engineering Economy class are shown. Using Excel, the best-fit straight lines are plotted through the points for each of the semesters shown in the chart. The cumulative class GPA for each of the semesters is also shown in the inset table in the lower right of the

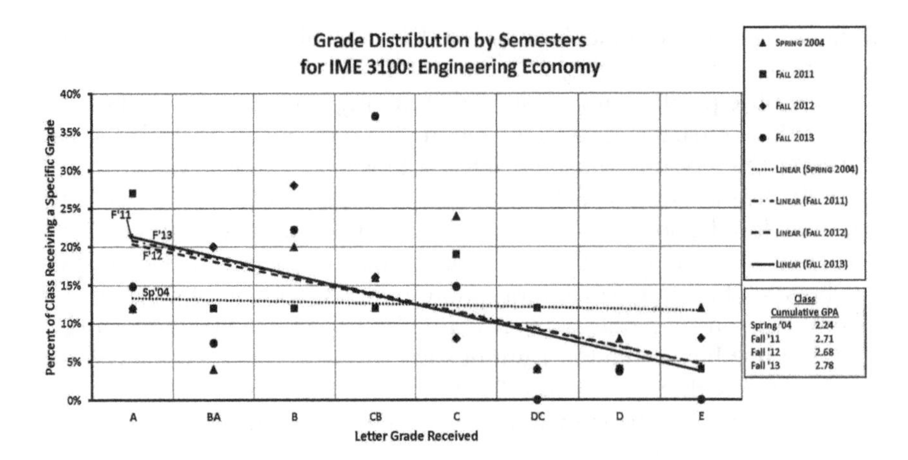

Figure 6.5 Grade distribution for selected semesters for Engineering Economy class.

graph. The spring 2004 (Sp'04) semester was chosen since this was the last semester when I taught this course as a traditional class. From fall 2004 onwards, I started making incremental changes in how I taught the course, finally teaching it in the hybrid classroom format (using flipped classes) for the first time in fall 2011. Hence, fall 2011 (F'11) was selected to see if there were any changes because of teaching the course as a hybrid class. Since then I have taught all offerings of the Engineering Economy course in the hybrid format. As such, I also selected fall 2012 (F'12) and fall 2013 (F'13) semesters to study the variation from semester to semester when teaching in the hybrid format.

The following conclusions can be drawn from the best-fit lines in Figure 6.5:

1. There is a significant difference in the grade distribution for the class between traditional teaching (Sp'04) and hybrid teaching (F'11 through F'13).
2. Hybrid teaching (F'11 through F'13) results in a substantially greater percentage of higher grades (A and BA) and a substantially smaller percentage of students failing the class (E) as compared to traditional teaching.
3. No significant variations were observed from semester to semester when the course was taught as a hybrid class (F'11, F'12, and F'13).
4. As shown in the inset table, the class cumulative GPA using traditional teaching (spring 2004) was 2.24 and using hybrid teaching it ranged from 2.68 to 2.78 (fall 2011 through fall 2013). This indicates an improvement in the class performance of half a letter grade (0.5) by switching from traditional teaching to hybrid teaching.

Table 6.1 Grade distribution by semesters and confidence intervals for the difference in proportions

		Course grade at end of semester							
		A	BA	B	CB	C	DC	D	E
Semester	Sp'04	12	4	20	16	24	4	8	12
	F'11	27	12	12	12	19	12	4	4
Confidence interval	Lower	0.042	0.006	−0.181	−0.136	−0.164	0.006	−0.106	−0.154
	Upper	0.258	0.154	0.021	0.056	0.064	0.154	0.026	−0.006

To validate this observation, I performed a statistical analysis of the difference between the proportions of students in a grade range between spring 2004 and fall 2011 semesters, as shown in Table 6.1. The results in the table demonstrate that the change in teaching methodologies from traditional to hybrid yield statistical differences in the highest (A and BA) and the lowest (E) grade groups (shown with shaded boxes in the Table) where the 95% confidence intervals for the differences of proportions are (0.042, 0.258), (0.006, 0.154), and (−0.154, −0.006), respectively. This validates the conclusion that with hybrid teaching using the flipped classroom approach the students perform better and fewer students are falling behind and failing the class.

6.11 Benefits with the flipped classroom approach to learning

In this case study, I have discussed my approach to using technology in teaching courses using the flipped classroom and hybrid and online learning. In the previous section, I analyzed and compared the results showing the benefits which I have achieved by changing my teaching methodology. Now let me discuss the benefits which have been obtained by my students when using these approaches to teaching.

1. The course is presented in a format that today's generation enjoys. Since they have grown up in the digital age, they enjoy using the computer to learn. My methods give them versatility since they can view my lecture videos on the computer, tablet, or smart phone.
2. The flipped classroom to some extent is like online learning since the student must do all the work themself. However, in the hybrid method the student has the opportunity of meeting with the instructor on a regular basis to clarify any materials with which they may be having difficulty.

3. Since I have limited the length of my video lectures to about 15 min, a student can complete viewing each video in a single sitting without much distraction from other sources such as social media.
4. Since my course plan is structured around the week, it gives greater flexibility to students. Based on their class schedule, part-time work schedule, and the individual's most productive time (morning, afternoon, evening, or night), the student can get to the course materials at a time of their choosing without affecting their other activities if it is completed sometime during the week.
5. Since the lectures are recorded on videos, the student can get a better understanding of the knowledge base. They can "master" each topic by reviewing the video or its specific parts multiple times, if necessary, before moving on to the next topic.
6. Each student can work at their own pace. Because of the differences in the capability of grasping materials, a student can spend less time or more time on individual topics in the course on an as needed basis. The flipped classroom method gives flexibility to the student to use as much time as needed. They can also use the online materials to review before an exam.
7. The students can perform better when taught using the flipped classroom method. Since this method allows greater flexibility, they can increase the "mastering" of the subject resulting in their getting a better grade in the course. Also, by having an improved understanding of the material in the course, they can be expected to perform better in other courses where this knowledge base is required.
8. The flipped classroom method of learning gives each student an equal opportunity to do well if they have such a desire. Hence, it puts an increased responsibility upon the student.
9. By being taught using the flipped classroom method, the student learns how to self-study. This will prove to be beneficial to them in the future as they pursue lifelong learning. In today's fast-changing world of technology, the knowledge acquired for the Baccalaureate degree which engineers receive will only take them through a few years of their working career before their education is outdated. Because of their geographical location, they may be restricted in acquiring additional coursework at an educational institution while working full time and will have to resort to self-learning using the Internet and other forms of technology.

I have discussed several benefits which the student can expect to obtain through the flipped classroom and hybrid teaching. However, for an instructor who may be debating whether they should make a move toward investing their time to prepare their course for flipped classroom offering, I have the following suggestions, which are based on my personal experiences.

1. Remember the saying "Rome wasn't built in a day." Don't try and convert your course in a single semester. Undoubtedly, it is a lot of work. Proceed with this like a project spaced out over 2–3 years and handle conversion of a part of the course at any time.
2. As you convert one segment of the course, seek input from your students on their experiences with the original segment and the converted segment. Use this feedback as you make additional changes.
3. If you are currently not using eLearning, start using some of its features in the first semester and gradually add on more features in subsequent semesters.
4. Many people "freak out" when they think about using technology in their work. Remember that technology is not a "monster" and is only there to help you in making your job easier. Approach it with a positive attitude and learn using a small component at a time. Before long, you will become an "expert" in using it.
5. As you delve into "new" areas, you cannot do it alone. As such, never be afraid to ask those around you. There is no sense in "reinventing the wheel." Use the experiences of others—it will be much easier on you and you can accomplish what you set out to do much faster.
6. Remember, you will be the one to reap the benefits from your additional time investments. Since you will probably be teaching the same course semester after semester, once you switch to hybrid learning, you do not have to lecture on the same materials each semester. Once you have recorded your lecture videos, you can reuse them for several semesters, maybe with minor modifications done occasionally since the knowledge base does not change that rapidly. If you teach the course as a hybrid class, you don't need to meet the class that frequently. This gives you more "free-time" to pursue other activities.

6.12 Conclusion

There is no doubt that my innovations in the design and presentation of the Engineering Economy course have brought the delivery of this course, and learning in general, to a whole new level. I have moved away from the traditional methods of student note-taking and testing and made the course "paperless." I have developed a procedure to record the lectures onto my laptop without the help of any other person or equipment. Instead of devoting all my lectures in class on the course material, I am using the "flipped" classroom model and streaming my "lecture videos" on eLearning so that students can view them outside the classroom at their convenience. I now use the scheduled class meetings to administer quizzes and exams and for solving problems. I am taking learning to a whole new level by applying the learned materials to solve problems. My analysis of the grade distribution in the class comparing the two methods of teaching indicates that my new format (hybrid class using the flipped classroom approach) in teaching the course is better than the traditional methods.

In the "flipped classroom" model of education, routine activities such as knowledge and information dissemination are done at home and the testing and application of the materials such as problem-solving is moved to the classroom. In the "online" model, both the information dissemination and the application are done at home. However, in this model, due to minimal or no interactions between the student and the instructor, the level of applying the material is somewhat limited. I have developed a combination of both these models where the course is offered as a "hybrid" model. Online testing has always been a concern to the "traditional" instructors because it encourages plagiarism, especially if the course is offered simultaneously to students in a limited geographical area such as on-campus. In my hybrid offering of the course, the information dissemination and its application are done at home and all the testing is done in a supervised environment in the classroom. My hybrid model also provides students an opportunity to discuss any problems in the application of the material with the instructor in the classroom. As a result, although the student–instructor contact hours have been reduced significantly, the students are being challenged at a higher level of learning and, on an overall basis, are performing significantly better.

By incorporating new and emerging technologies in my teaching, these changes have fostered critical thinking among my students and inspired them to apply the techniques learned in the course to real-life problems. The results presented clearly demonstrate the benefits achieved by implementing the various techniques described. Although these techniques were applied in teaching the Engineering Economics course, most of them can easily be adapted to other courses in engineering and other fields. These techniques are so simple that they have a potential for widespread adoption, especially since most academic institutions today already have an eLearning platform available. All it will require is the desire to make the change and a time commitment by the faculty member.

References

1. Bart, M. The Five R's of Engaging Millennial Students, Available: www.facultyfocus.com/articles/teaching-and-learning/the-five-rs-of-engaging-millennial-students/, December 21, 2013. (Posted 11-16-11).
2. Lage, M., Platt, G., and Treglia, M. Inverting the classroom: A gateway to creating an inclusive learning environment, *The Journal of Economic Education*, vol. 31, no.1, pp. 30–43, 2000.

3. Lage, M. and Platt, G., The internet and the inverted classroom, *The Journal of Economic Education*, vol. 31, no. 1, p. 11, 2000.
4. Saenz, A. Is the Khan academy: The future of education? (video), Available: http://singularityhub.com/ 2010/09/11/is-the-khan-academy-the-future-of-education-video/, December 22, 2013. (Posted 9-11-10).
5. Kolowich, S. Exploding the lecture, Available: www.insidehighered.com/news/2011/11/15/professor-tries-improving-lectures-removing-them-class#ixzz1lFXwcJMr, December 23, 2013.

chapter seven

RLaaS-Frame

A new cloud-based framework for remote laboratory system rapid deployment

Xuemin Chen
Texas Southern University

Qianlong Lan, Ning Wang, and Gangbing Song
University of Houston

Hamid R. Parsaei
Texas A&M University at the Qatar

Contents

In this chapter, a Remote Laboratory as a Service (RLaaS) is defined as a new standardized cloud-computing model. Based on the RLaaS model, a new cloud-based framework, namely RLaaS-Frame, is proposed to flexibly and rapidly deploy the remote laboratory system. To implement the RLaaS-Frame, a Wiki-based remote laboratory platform and a mobile-optimized remote laboratory application framework are successfully integrated into the RLaaS-Frame to support cloud-based remote laboratory system implementation. To address the stability of RLaaS-Frame-based remote laboratory system, Docker is used to be a container for imaging all components of the RLaaS-Frame. To illustrate the effectiveness of the RLaaS-Frame, a remote Smart Vibration Platform (SVP) is revamped. As a standardized framework, the RLaaS-Frame can be extended to other disciplines, such as physics, chemistry, biology, etc. This RLaaS-Frame will accelerate the adoption of remote laboratory technology and benefit online education, academic research, and industrial applications.

7.1 Introduction

With the dramatic advancements of information technology in recent decades, an increasing number of remote laboratory systems and innovative concepts for remote experimentation have emerged [1]. Many research teams have worked in this field and state-of-the-art technologies also have been developed and applied to support more complex remote laboratory systems including iLab [2], WebLab-Deusto [3], the Distance Internet-Based Embedded System Experimental Laboratory (DIESEL) [4], etc. A generalized remote laboratory system is based on the Browser/Server (B/S) architecture [5,6] with three modules normally: a client web application in user's devices, a middleware which is usually a learning management system (LMS) in server, and an experimental equipment control application in workstation. However, a common denominator of most existing remote laboratory systems, whether for academic or for industrial purposes, is that they offer standalone solutions with limited or no capability to cooperate with other platforms [7,8]. To easily sharing of remote laboratory experiments on large-scale incompatible platforms, a unified, standardized, and reliable software architecture is crucial. These critical issues have been pointed out by research works in [7–9], and can be listed as follows:

1. How to rapidly distribute and establish the remote laboratory system architecture on a global scale?
2. How to efficiently and reliably duplicate the physical components and environment of remote laboratory systems at a new location?
3. How to deliver the remote experimental services to the end users in a short period of time?

The revolutionary progress of cloud-computing technology provides an effective solution for these essential issues. Recently, cloud computing has become increasingly popular due to its unique advantages over traditional computing models [10]. The National Institute of Standards and Technology (NIST) makes a comprehensive description in identifying cloud-computing technology with the parameters [11]. Based on the definition of cloud-computing model from NIST, the Software as a Service (SaaS) is the first layer of standard cloud. The Platform as a Service (PaaS) is the second layer of standard cloud, and the Infrastructure as a Service (IaaS) is the third layer of standard cloud [11]. These three layers provide three different series of services to the various users.

Cloud computing adopts concepts from Service-Oriented Architecture (SOA), and provides a new approach to supporting the system developers to break all of the functional modules of the system into services [12]. Cloud-computing model incorporates the well-established standards and best practices, which are gained in the domain of SOA, to allow global and easy access to cloud services in a standardized way [13,14]. Web Service, which has a well-defined set of implementation approaches, is the most typical example of an SOA. However, the real-time data communication is still a critical challenging for cloud-based systems, and it has been an active area of cloud-computing research [15,16]. To develop a suitable cloud-based solution for flexibly and effectively deploying the existing remote laboratory system, several current research activities are summarized as follows.

Case 1: A research team proposed a Laboratory as a Service (LaaS) Model for developing and implementing remote laboratory system at Spanish University for Distance Education (UNED) [17,18]. They put forward a general cloud-based remote laboratory architecture for distributed LMS, and used the Java plug-in and Java virtual machine (JVM) to tackle the real-time data transmission issue. However, two essential issues still cannot be solved: (1) there is not a standard definition of LaaS Model for other users. It is hard to be used for future remote laboratory rapid and flexible deployment; (2) there is still a plug-in issue for the web remote laboratory application to support real-time data transmission.

Case 2: A Cloud-based Remote Virtual Prototyping Platform was developed at Karlsruhe Institute of Technology (KIT) [19]. They presented a cloud-based architecture allowing the design of virtual platforms and prototyping of the system including sophisticated software with prerecorded data or test benches. However, this architecture is not a standard definition of cloud-computing model. It is only a special cloud-based platform for the design and test of Multi-Processor system on chips instead, which makes it hard for future remote laboratory efficient and flexible deployment as well. As a special cloud-based platform, it has a restricted solution to support real-time data transmission.

Moreover, to expand the scope of remote laboratory system application into online education, the concept of Massive Open Online Labs (MOOL) was presented in Lowe's paper [20] and Salzmann et al. paper [21]. Their approaches are to deploy remote laboratories within Massive Open Online Courses (MOOC) software systems and infrastructures. However, it provides no standard cloud-computing model for efficiently implementing remote laboratory system in future.

Therefore, how to design a standardized cloud-based framework to rapidly and flexibly deploy remote laboratory system still have not been resolved very well so far [17–20]. To address this vital issue, a new cloud-based RLaaS framework is proposed and designed in this chapter. Remote laboratory systems building on a unified and standardized RLaaS-Frame can be easily deployed, and all resources can be shared and accessed by users with great convenience. Consequently, the RLaaS-Frame will significantly simplify works needed to integrate and implement remote laboratory systems on a large scale.

To demonstrate the feasibility of the RLaaS-Frame, a Wiki-based remote laboratory platform and the mobile-optimized application framework have been integrated into the RLaaS-Frame. A revamped cloud-based remote SVP experiment application has been implemented [22]. Meanwhile, this novel framework also can be applied for the integration of industrial equipment remote control and monitoring applications. It will be a significant improvement for the remote laboratory development technology in future.

7.2 *Previous works*

In our previous works, to integrate existing experiments and laboratories in different places into a Browser–Server architecture software infrastructure, a global remote experiment scheduler and federated authentication solution [23] were designed and implemented. Furthermore, a unified framework was also proposed by the authors which solved several critical issues and had the ability to provide real-time video and real-time data transmission without software plugins and firewall issues [24,25]. A novel Wiki-based remote laboratory platform [26] also has been developed to offer a more flexible tool for collaborative learning. Moreover, to integrate the remote laboratory applications into the mobile environment for Mobile Learning (M-Learning), a mobile-optimized remote laboratory application architecture [27] has been developed as well. To select an efficient and stable approach for cloud-computing system implementation, the comparison of cloud-computing platform implementation approaches is summarized in this section as well.

7.2.1 *The global remote laboratory architecture based on the unified framework*

To build up the global remote laboratory as shown in Figure 7.1, three servers are set up at three universities, i.e., Texas A&M University at Qatar (TAMUQ), University of Houston (UH), and Texas Southern University (TSU). All of three servers are based on the Linux/Node.js/Apache/MySQL/PHP (LNAMP) system architecture. A unified framework based on the Web 2.0 technology has been designed and developed [28]. The kernel of this framework includes three modules: client web application, server application, and experimental control application. The server application is based on Web-Service technology and directly built on top of MySQL database, Apache web server engine, and Node.js web server engine [29,30]. For the server application design and implementation, there are three vital technologies: the Socket.IO, Node-HTTP-Proxy, and HTTP Live Streaming (HLS) protocol. Node-HTTP-Proxy is used for the novel video transmission

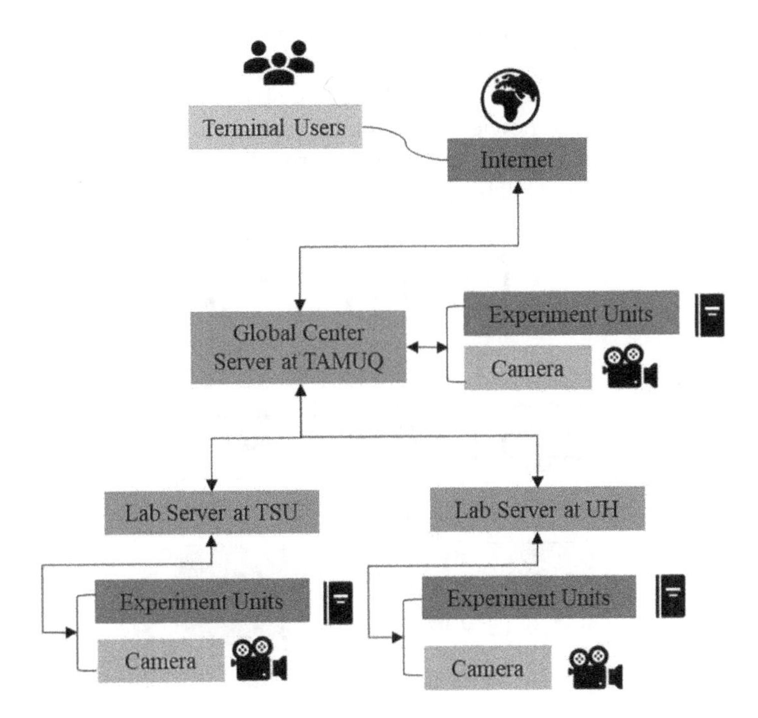

Figure 7.1 The global remote laboratory architecture.

approach which is based on HLS protocol for real-time system monitoring [24], and experiment data and control commands transmission and traversing firewall [25]. The server application runs on Linux CentOS server. Currently, the experiment control application is based on the LabVIEW and uses Socket.IO for real-time communication with the server application. Moreover, the experiment control application solution is designed to use the NI LabVIEW for implementing the experiment hardware platform control as well. All of the experiment control applications run on the workstations with Windows OS. The client web application is designed and implemented based on Hyper Text Markup Language (HTML), Cascading Style Sheets (CSS), and JQuery/JQuery-Mobile JavaScript libraries. The server-based Mashup technology is applied for user interface (UI) integration. The client web application can run on most of the current browsers such as Microsoft Edge, Firefox, Chrome, Safari, etc. The UI can run in any terminal devices without the installation of any plug-in or software except the Java runtime engine for IE which is required for real-time video streaming.

7.2.2 A Wiki-based remote laboratory platform

To utilize advantages of the Wiki technology [31] and remote laboratory technology for engineering online education, a collaborative and cooperative learning environment supported by the remote laboratory technology has been designed and implemented successfully [26]. With this Wiki-based remote laboratory platform, students and instructors can collaboratively design and implement new experiments to support student-centered engineering online learning. As shown in Figure 7.2, there are three layers in this platform: (1) The Database layer: it includes a Data Pool (DP), which

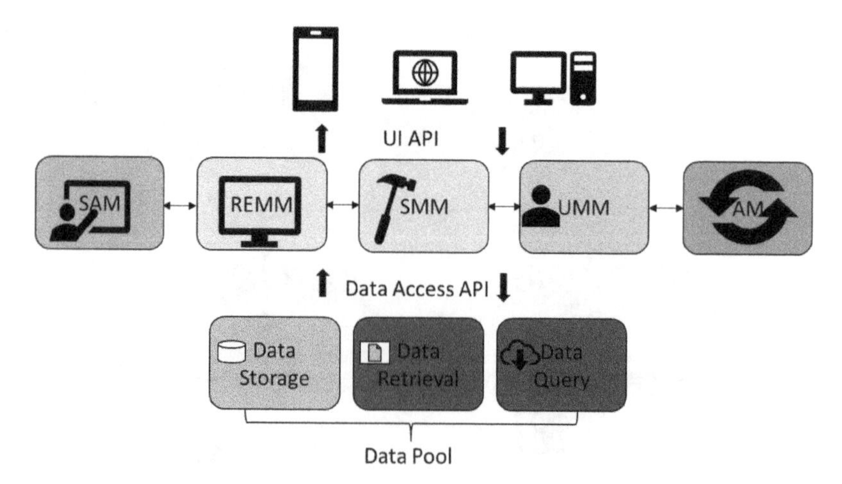

Figure 7.2 The architecture of Wiki-based remote laboratory platform.

provides data storage, data retrieval, and query to support platform layer. (2) The Platform layer: it is the core of the novel Wiki-based remote laboratory platform. It includes the following modules: a Study Aids Module (SAM), a Remote Experiments Management Module (REMM), a System Management Module (SMM), a User Management Module (UMM) and an Appending Module (AM). To support the student-center collaborative and cooperative learning, three modules, SAM, REMM, and SMM, work together to provide some new essential functions. The AM provides a social context for students learning. Moreover, the UMM provides the user management functions to support the standard usage of the systems. (3) The Client layer: it includes a set of User Interface Application Programming Interfaces (UIAPIs) to support the different UIs for students' various learning activities.

7.2.3 A mobile-optimized remote laboratory application architecture

For supporting M-Learning, a mobile-optimized architecture has been developed for mobile remote laboratory application development [27]. As shown in Figure 7.3, the mobile-optimized remote laboratory application architecture includes two layers: optimized application layer and unified framework layer. With the Ionic Framework [32], the mobile-optimized application layer can support applications to run on most of the popular

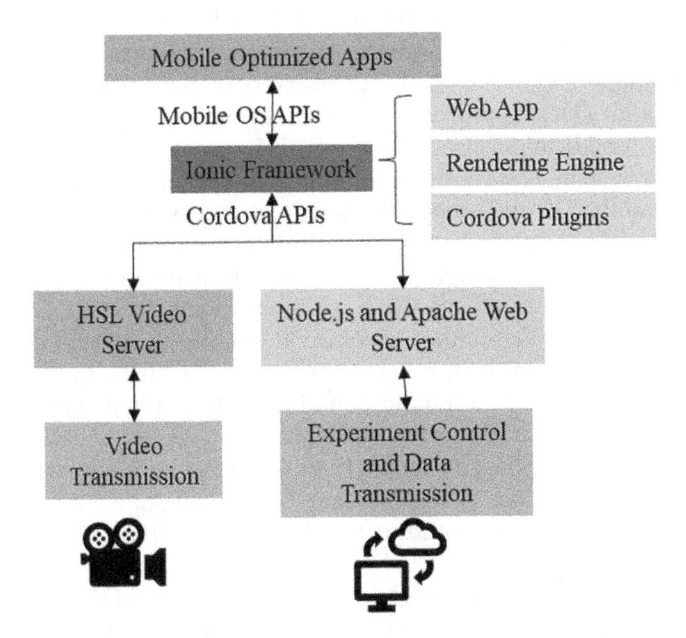

Figure 7.3 The mobile-optimized remote laboratory application architecture.

mobile platforms. To integrate the remote laboratory technology into the new mobile-optimized application architecture, the unified framework [28,33] has been integrated into this mobile remote laboratory application framework. As the new mobile-optimized application architecture integrated to take the advantages of both native mobile application and web application, two essential issues of the mobile web applications, i.e., the running performance issue and the hardware accessibility issue, have been improved significantly. Moreover, it also solved the cross-platform running issue of mobile native applications. It seamlessly combined the unified framework and Ionic framework together to deliver the excellent remote laboratory services for students' learning.

7.2.4 The deployment approach selection of cloud computing platform

Nowadays, cloud computing is a leading-edge technology. More and more deployment approaches of cloud have emerged recently [34]. The majority of cloud deployments are built on Virtual Machines (VMs). VMs are specific portions of the cloud infrastructure (e.g., processors, memory) selectively chosen and used to create the sense of a standalone computer [35]. However, currently, the Docker is increasable utilized for cloud deployment as it offers the significant benefits to the cloud users [36]. To select a suitable approach from VM or Docker, three primary factors should be considered [37,38]. These factors include: (1) the functional differences between VM and Docker; (2) the level of interdependence between private and public cloud components; (3) users' willingness to customize their cloud platform.

VM and Docker represent two different ways to create virtual resources for an application running. VM shares only hardware with Operating System (OS). With a container, virtualization takes place at the OS level, so the OS and some middleware are possibly shared. From the functional point of view, VM is more flexible, because the guest environment where the applications run is similar to a bare-metal server. The users' special OS and middleware are independent from the VM on the same server. With a container, a common OS and middleware elements need to be accommodated, because each container uses the core server platform and shares it with other containers. Docker is usually deployed through management platforms [39,40]. It is also generally easier to operationalize container-based clouds than VM-based clouds, where management tools are more varied. Docker is an open source engine, and it primarily focuses on automating the deployment of applications inside software containers and OS-level virtualization on Linux. Table 7.1 shows the main differences between VM and Docker. Figure 7.4 shows the result from the Google Trend Chart [41]. Comparing with VM, the number of searches for Docker has steadily risen since its initial release in early 2013.

Table 7.1 Comparison of VM and Docker

Virtual machine	Docker
Represent hardware-level virtualization	Represent OS-level virtualization
Heavy weight	Light weight
Slow provisioning	Real-time provisioning and scalability
Limited performance	Native performance
Full isolated and hence more secure	Process-level isolation and hence less secure
Minutes level boot time	Seconds level boot time

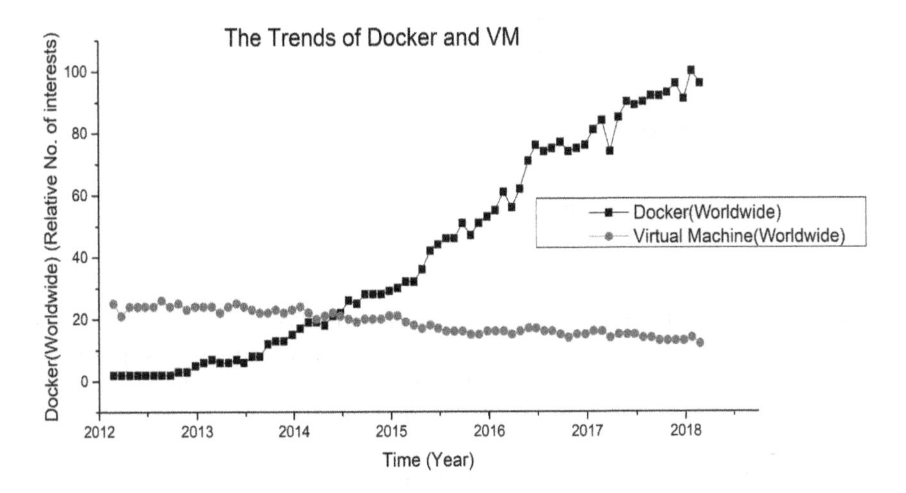

Figure 7.4 The technology trends of virtual machines and Docker.

As shown in Figure 7.5, VMs and Docker differ on quite a few dimensions [42]. However, primarily because Docker provides a way to virtualize an OS for multiple workloads to run on a single OS instance. With VMs, the hardware is virtualized to run on multiple OS instances. Docker doesn't require or include a separate OS, and it relies on the Linux kernel's functionality and uses resource isolation. Docker's speed, agility, and portability make it yet another tool to help streamline software development. In general, users can gain all the benefits of the container in the cloud deployments. As Docker is based on Linux containers, Docker-based cloud is the best strategy for businesses with standardized OS and middleware. Based on our unified framework architecture [28] and Wiki-based remote laboratory platform [26], which have been developed on Centos OS, the Docker should be the more suitable selection for the RLaaS-framework implementation.

In order to combine the advantages of VM and Docker, an optimized solution to support the RLaaS-Frame is proposed. OpenStack, as a very popular software platform, is used in our optimized solution. OpenStack

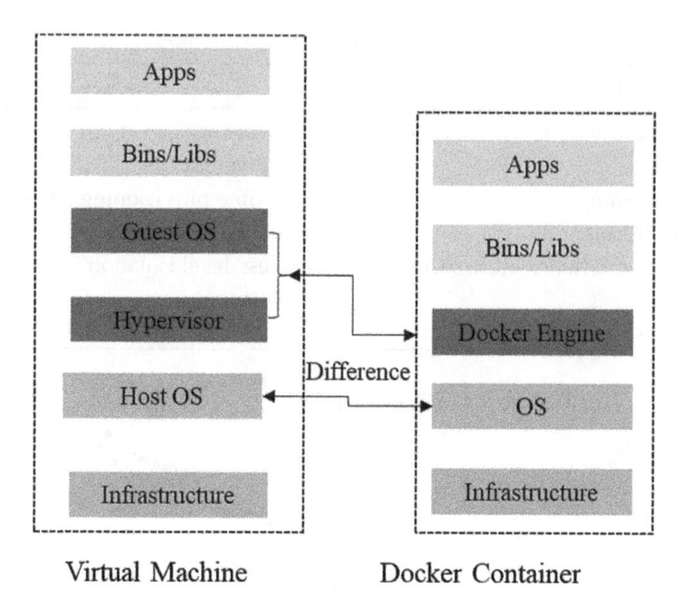

Figure 7.5 Comparison of virtual machines and Docket container.

is a free cloud platform for private and public clouds. It supports an IaaS cloud and provides allocation for computational resources in support of both interactive and computationally intensive applications [43]. The software platform consists of interrelated components that control diverse, multi-vendor hardware polls of processing, storage, and networking resources throughout a data center. The key components of OpenStack include Nova Computing module, Glance Image Service, Swift Object Storage, Heat Orchestration, and Neutron Networking, etc. [44]. Docker provides deterministic software packaging and fits nicely with the immutable infrastructure model such as the optimized mobile application architecture and Wiki-based remote laboratory platform. OpenStack offers a complete data center management solution in which containers or hypervisors are only part of an RLaaS-Frame. OpenStack also includes multi-tenant security and isolation, management and monitoring, storage and networking and more. All of these services are needed for RLaaS-Frame to manage the cloud or data center. So Docker on top of OpenStack serves is an excellent containerization of micro-services pods for the RLaaS-Frame.

7.3 Methodology

The RLaaS model originates from promoting the traditional remote laboratory technology to the new cloud-based remote laboratory technology. As shown in Figure 7.6, the RLaaS model architecture is divided into three-part services, experiment Application (experiment applications) part services,

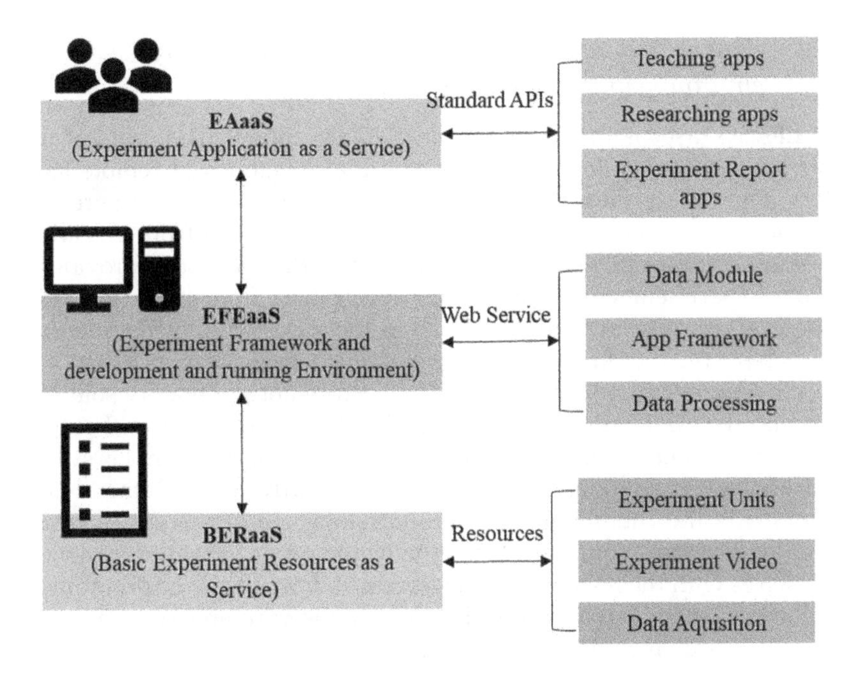

Figure 7.6 Remote Laboratory as a Service (RLaaS) model architecture.

platform (experiment development and integration platform) part services, and resources (experiment data, experiment video, etc.) part services.

7.3.1 Experiment applications as a service (EAaaS) layer

The EAaaS layer is the first layer of the RLaaS model. The users of this layer are those who use the remote laboratory applications for education and research. To meet the needs of the different users, the RLaaS model offers the different UIs through this layer. To simplify the users' operations and reduce the efforts and cost of the system maintenance, we adopt the Browser–Server architecture applications to provide the EAaaS layer service to the end users. It supports most of the popular web browsers (such as Chrome, IE, Safari, Mercury) which are running in different terminal devices (desktops, laptops, mobile phones, tablets, etc.). The goal of our design for the EAaaS layer is to provide the flexible services to the different end users through the different remote laboratory applications.

Experiment Applications mainly include the following:

- Remote laboratory applications (RLaaS-based application) for education.
- Remote laboratory applications (RLaaS-based application) for research.
- Other applications (RLaaS-based application)

7.3.2 *Experiment development framework and running environment as a service (EFEaaS) layer*

The EFEaaS layer is the second layer of RLaaS model. The users of this layer are those who design and develop their customized remote laboratory applications running in web browsers. They will gain the remote laboratory development and integration support services from this layer.

From the RLaaS model architecture, the RLaaS model provides a remote experiment development and integration framework and platform level running environment (such as the series of APIs, the interface of Database, the interface of experiment data analysis component and tool kits, etc.). Users can develop their customized remote laboratory solutions based on the APIs which are provided by the EFEaaS layer without too much effort expended in developing and maintaining the complex system software layers and managing the underlying hardware. With the EFEaaS layers' offers, the underlying hardware and storage resources automatically match applications requirements internally, and users don't need to concern themselves with the resources management and maintenance procedures.

Experiment integration/development framework and running environment mainly include the following:

- Experiment application framework for secondary development
- Experiment data transmission protocol
- Experiment data analysis method and algorithm component
- Experiment data processing tool kits

7.3.3 *Basic experiment resources as a service (BERaaS) layer*

The BERaaS layer is the third layer of RLaaS model. The users of this layer are those who request experiment equip mental setup and configuration services, basic experiment resources, and environment delivery services for their customized remote laboratory system.

From the RLaaS model architecture, the BERaaS layer offers the Basic Experiment Unit (OS, VM, software for experiment data collection, experiment server computers configuration, network cameras configuration, experiment equipment configuration, etc.), experiment equipment control software components, and experiments data (experiment data, experiment videos, etc.). The BERaaS layer also includes the Docker open source application virtualization technology. Docker utilizes the extra storage space in servers and data centers by RLaaS-Frame architecture definition.

Basic Experiment Resource mainly includes the following:

- Experiment unit cluster management
- Experiment unit duplication sharing

- Experiment environment setup and configuration image duplication
- Experiment equipment control component
- Collecting and storage experiment data
- Experiment video

Docker is a suitable tool of managing multiple containers on a single server for RLaaS-Frame implementation. OpenStack Nova Computing module makes Docker much more powerful as the containers. OpenStack Nova computing module can manage many severs which can deploy hundreds of containers. Docker can be integrated into OpenStack Nova as a form of the hypervisor (Containers used as VMs). Docker's driver is a hypervisor driver for OpenStack Nova Computing module, so it's a lightweight solution for RLaaS-Frame to utilize the OpenStack open source service.

The Docker and OpenStack Communication Model shows in Figure 7.7. The Nova driver embeds a tiny HTTP client which talks with the Docker internal Rest API through a Linux socket. It uses the HTTP APIs to control containers and fetch information about them. The driver will fetch images from the OpenStack Image Service (Glance) and load them into the Docker filesystem. Images may be placed in Glance by exporting them from Docker. In this way Docker brings some unique benefits for RLaaS-Frame as follows.

- Process-level API: Docker can collect the standard outputs and inputs of the process running in each container for logging or direct interaction; it allows blocking on a container until it exits, setting its environment, and other process-oriented primitives which don't fit well in libvirt's abstraction.
- Advanced change control at the filesystem level: Every change made on the filesystem is managed through a set of layers which can be snapshotted, rolled back, etc.

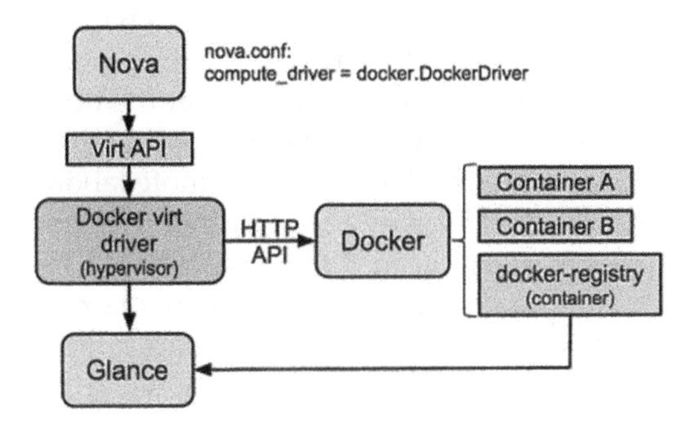

Figure 7.7 Docker and OpenStack Communication Model.

- Image portability: The state of any Docker container can be optionally committed as an image and shared through a central image registry. Docker images are designed to be portable across infrastructures, so they are a great building block for hybrid cloud scenarios.
- Build facility: Docker can automate the assembly of a container from an application's source code. This gives developers an easy way to deploy payloads to an OpenStack cluster as part of their development workflow.

7.3.4 Characteristic of the RLaaS-fame

As the RLaaS model is based on the cloud-computing technology, it not only comprises certain standard characteristics from the cloud-computing technology but also has some of its unique special characteristics. The specific characteristics of the RLaaS model are illustrated as follows:

- The standard APIs (Application Programming Interface) of RLaaS model accessibility to software that enables machines to interact with RLaaS-based platform in the same way as the UI facilitating interaction between users and computers. The RLaaS model consists of a set of standard APIs with REST (Representational State Transfer) based on SOA Web-Service architecture. The standard APIs are mainly utilized to support the cloud-based applications development and integration.
- Device and location independence enable users to access systems only using a web browser regardless of their locations or what devices (desktop, laptop, tablet, pad, mobile phone, etc.) they are using. As Infrastructure is off-site (provided by different physical laboratories in the different places in the worldwide) and accessed via the Internet, users can connect with the RLaaS-based remote laboratory platform from anywhere in anytime.
- Virtualization technology allows for easily and smoothly sharing and utilization of servers and storage devices which are in the physical laboratories in different locations. RLaaS-based applications can be easily migrated from one physical server to another (even though they are in different locations in the worldwide scale).
- Security performance in RLaaS-based remote laboratory platform is often better than, or at least as good as other traditional remote laboratory systems, in part because the RLaaS-based remote laboratory service provider can centralize management of experiment data and increase security-focused experiment resources. Meanwhile, the service provider can also establish the unified security passcode in the standard experiment data transmission protocol which is developed in RLaaS model to control the experiment data security. However,

the complexity of security is greatly increased when experiment data are distributed over a wider area or greater number of devices. Consequently, the security management module of RLaaS model needs to be constantly upgraded based on the solutions to different specific security issues.

- Maintenance and update of the RLaaS-based platform become easier. As the end users will only use web browsers without any software plug-in to access the RLaaS-based remote laboratory platform from anywhere in the world, they do not need to worry about software update issues. The RLaaS-based platform only needs to perform maintenance on the server side.

7.4 Implementation of the proposed RLaaS-Frame

To answer our research question, *How to design a standardized framework to rapid and flexibly deploy remote laboratory system?*, A new cloud-based framework, namely RLaaS-Frame, for remote laboratory system rapid deployment is proposed. RLaaS-Frame is a web application framework and an SOA built upon the cloud-computing technology. The using of real laboratory resources (experiment equipment, experiment data, experiment control software, experiment methods, etc.) are delivered with the standard protocol as a service over a network (typically the Internet network). The end users use the remote laboratory services without the need to understand how the system and infrastructure work with the new RLaaS-Frame-based remote laboratory system.

7.4.1 The Implementation of new RLaaS-Frame

Base on the RLaaS model architecture, the new RLaaS-Frame is designed and implemented to provide the remote laboratory services for end users. Table 7.2 shows the features of the RLaaS-Frame, and it includes two parts: experiment applications and platform services as shown in Figure 7.8.

7.4.1.1 Experiment applications

Based on the EAaaS layer of RLaaS model, there are many different remote experiment applications to provide the different remote laboratory services for users' academic and research activities. The experiment applications can run on most of the popular browsers and mobile platforms, and are built on HTML, CSS, PHP, AJAX and JQuery/JQuery-Mobile JavaScript libraries. Moreover, an adapter API layer for data exchange is required for data communication between the experiment application layer and the platform layer. To address the real-time data communication challenge of cloud-based system, Socket.IO protocol is used to implement the real-time experiment data transmission between the UI layer and the platform layer. Socket.

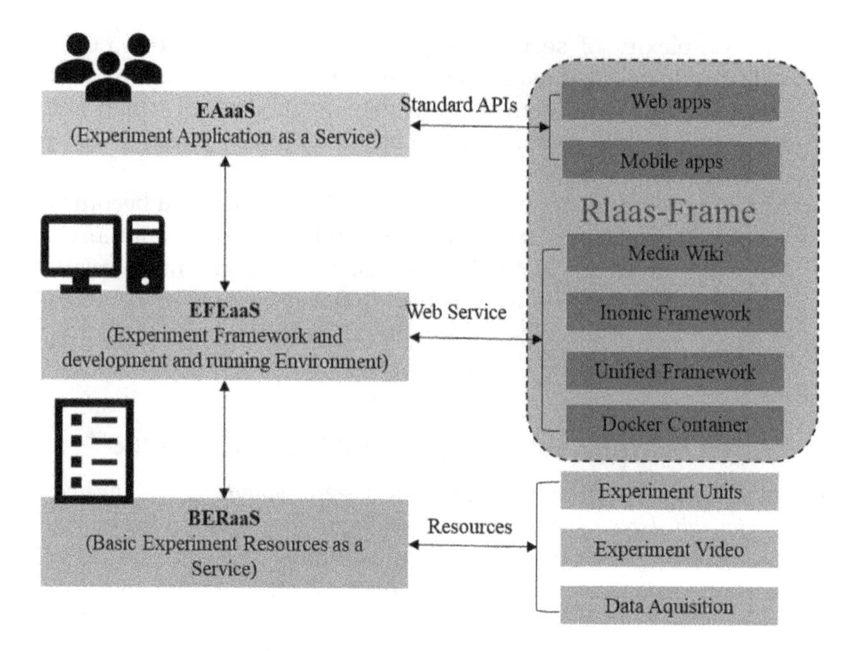

Figure 7.8 The RLaaS-Frame based on the RLaaS model definition.

IO is designed based on WebSocket, and enhances the WebSocket by providing built-in multiplexing, horizontal scalability, automatic JSON encoding/decoding, and more. Meanwhile, Socket.IO supports the real-time web applications in any popular browser. Socket.IO includes a client-side library for the browsers, and a server-side library supported by Node.js.

7.4.1.2 *Platform services*

To integrate the Wiki-based remote laboratory platform and mobile-optimized remote laboratory application architecture into the RLaaS-Frame, the Docker is adopted to provide a running environment for these frameworks. As shown in Figure 7.8, all of modules in the RLaaS-Frame, which include real-time data transmission module, real-time video transmission module, security module, MediaWiki engine, Ionic framework and all of the supporting components for the RLaaS-Frame which include Apache, Node.js, MySQL, PHP, RESTful Services, AngularJS are installed into the Docker image. The Docker container is created from Docker image, and it supports rapid deployment. Docker container is supported by Docker Engine, and it needs to be installed and configured in Linux OS. In the server side, Docker 0.9 is executed in the server OS, CentOS. To successfully execute Docker 0.9, the execution environment must be built up and configured. Some important build-in execution drivers, such as libcontainer, Libvirt, LXC, Systemd-nspawn, must be installed.

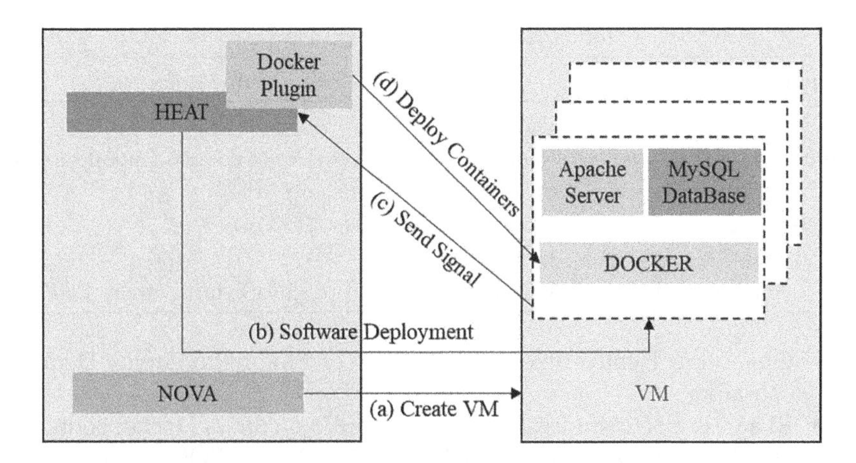

Figure 7.9 Deploy Docker containers with OpenStack.

Figure 7.9 shows the Docker Engine's execution environment. As the Docker can image the full-fledged RLaaS-Frame based server, it greatly reduces the burden of building, deploying and maintaining the entire remote laboratory systems, and only need to update the application layer based on users' various requirements.

7.4.1.3 Deploy Docker containers with OpenStack

To deploy Docker containers with OpenStack, there are two key components included: OpenStack Nova and OpenStack Heat. OpenStack Nova provides a platform on which OpenStack is going to run the guest machines. It's VM provisioning and management module that defines drivers that interact with underlying Docker containers. OpenStack Heat creates human and machine-accessible services for managing the entire lifecycle of infrastructure and applications within OpenStack clouds. It contains human readable template with simple instruction that is read by Heat Engine.

To integrate Docker with OpenStack, the OpenStack Heat provides the facility to create Docker containers within OpenStack clouds. Docker containers are implemented with RLaaS-Frame instances like Apache and MySQL. In this way, the components of RLaaS-Frame can easily integrate into OpenStack cloud-computing environment.

Figure 7.9 shows the communication between Heat, Nova, Docker, and the RLaas-Frame instance:

- OpenStack uses Nova to create VMs or hypervisors.
- OpenStack Heat deploys software packages and Docker into VMs.
- VMs will send signals to OpenStack Heat if the deployment cannot reach the Heat API

Table 7.2 Features of RLaaS-Frame

Properties	RLaaS-Frame
Services type	PaaS
Support for (Value offer)	Remote laboratory services (web apps/mobile apps)
User access interface	Web APIs and mobile app APIs
Virtualization	Application container (Docker)
Platform (OS & runtime)	Linux (CentOS)
Deployment model	Web apps/mobile apps (JavaScript/Python/PHP)

- OpenStack Heat utilizes Docker APIs or Plugins to deploy Docker containers.
- RLaaS-Frame instances package and build on top of Docker containers. The various services, including Apache, MySQL, frameworks, engines, etc., are implemented on Docker containers.
- RLaas-Frame instances utilize Docker to identify the required resources, edit the template, and integrate it on Heat.

7.4.2 Integrated a remote SVP experiment

To illustrate the effectiveness of the RLaaS-Frame, a remote SVP experiment was revamped based on the RLaaS-Frame as shown in Figure 7.10.

This mechanical engineering experiment has been incorporated as part of the remote laboratory series used in the Intelligent Structural Systems course at the UH. For the experiment integration, the detailed process is given below.

7.4.2.1 Experiment hardware setup

The SVP device, as shown in Figure 7.11, is assembled by using fabricated and purchased components. The SVP has a one-story flexible steel frame fixed on top of a plexi-glass box. In the plexi-glass box, there are electric circuit boards made to control the experiment. Other than the flexible steel frame, the SVP has a motor, SMA (Shape Memory Alloy) wires, and a magnetized iron clamped on a container of Magneto-Rheological (MR) fluid. The motor with an unbalancing weight is mounted on the top of the frame and connected to the driver from the box on the bottom. When the user controls the current going through the electrical circuit, the speed of the motor can be adjusted. The rotation of the motor leads the flexible frame to vibrate. Two SMA wires are hung across the frame. When the current goes through the wires, the temperature will increase. At a certain point of rising temperature, the SMA wires will shrink in length to reduce the vibration of the frame; this is called an SMA brace. A red steel tongue is placed downwards into the container of MR fluid. The magnetic iron clamped on the container can generate a magnetic field when it is

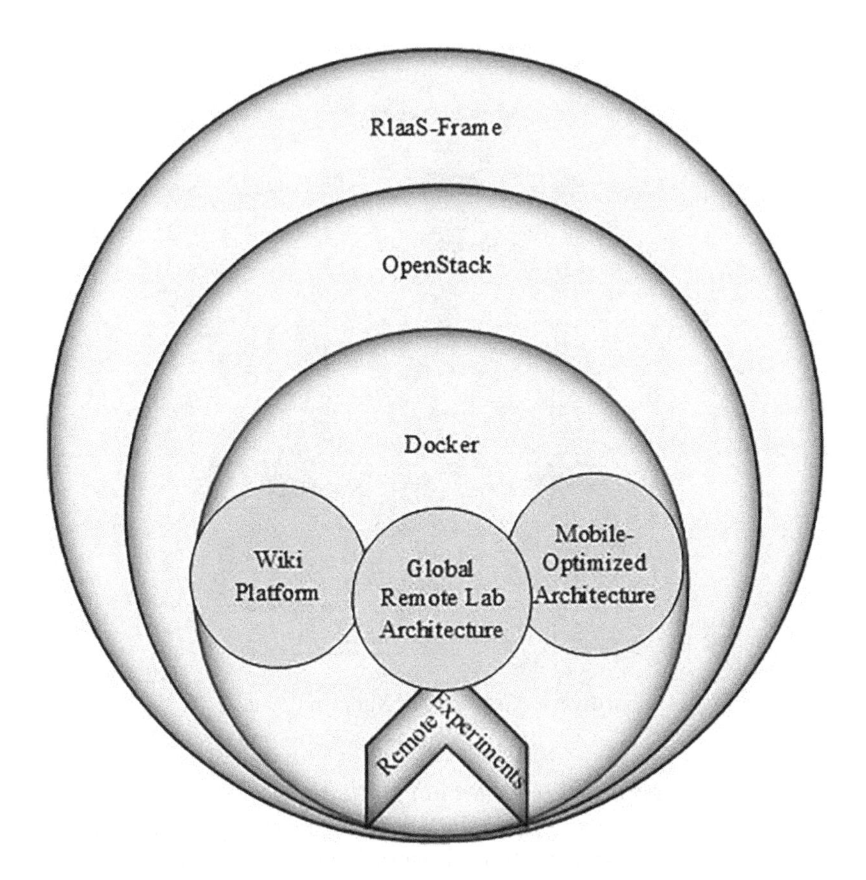

Figure 7.10 The RLaas-Frame environment.

turned on. The presence of the magnetic field increases the viscosity of MR fluid because MR fluid changes from fluid state to semi-fluid state under the magnetic field; this is called MR damper.

As a remote experiment, users can control the experiment remotely. To let users view the real-time response of experiment, a web camera was connected in the remote laboratory environment. The camera for the SVP is placed on a camera tripod in front of the experiment. A workstation is used to control the experiment, including the heat generation to the SMAs and the strength of the magnetic field. NI LabVIEW is installed on the workstation. Three DAQ 6008 USBs were connected to the workstation, and their voltage outputs and voltage inputs are controlled and sensed by LabVIEW program running on the workstation. The workstation is also connected to the web server via the network port. All data generated by LabVIEW is sent to the server and control commands are also sent from the server to the workstation via the Internet.

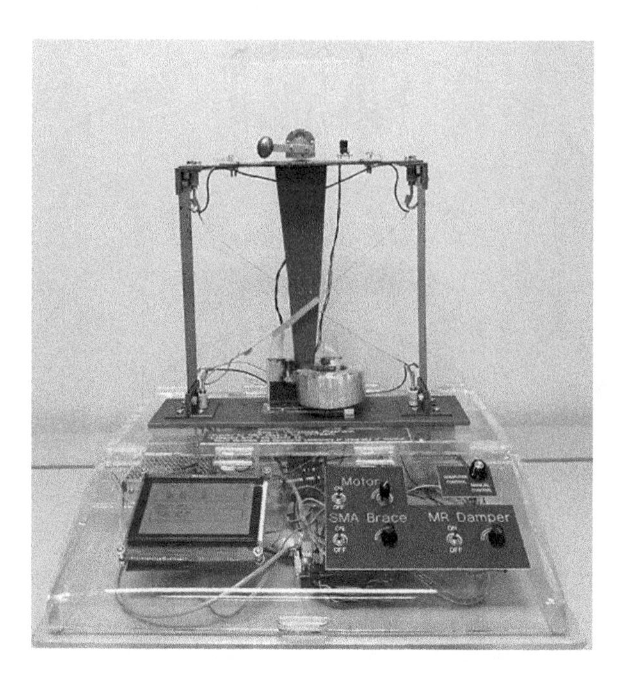

Figure 7.11 The SVP hardware. (Courtesy of Xuemin Chen.)

7.4.2.2 Experiment software integration

With the RLaaS-Frame, the remote SVP experiment application will provide the experiment services to end users. The software implementation of remote SVP experiment includes three tasks including the app implementation which includes web app and mobile app, platform integration, and experiment control application implementation.

a. *Application layer implementation.* As the RLaaS-Frame integrated the Wiki-based remote laboratory platform and mobile-optimized application architecture, the remote SVP experiment app provides the functions to support co-creation of experiment-based learning materials, users can work together to design and implement a new remote experiment UI based on their requirements. As shown in Figure 7.12, the UI of RLaaS-Frame-based remote laboratory platform includes five parts:
   - Real-time experiment video (in Control Panel),
   - Real-time experiment data display component (in Control Panel),
   - Real-time experiment control component (in Control Panel)
   - The experiment tutorial (in Tutorial)
   - The practices questions, reference contents, and quiz (in Functions Tab).

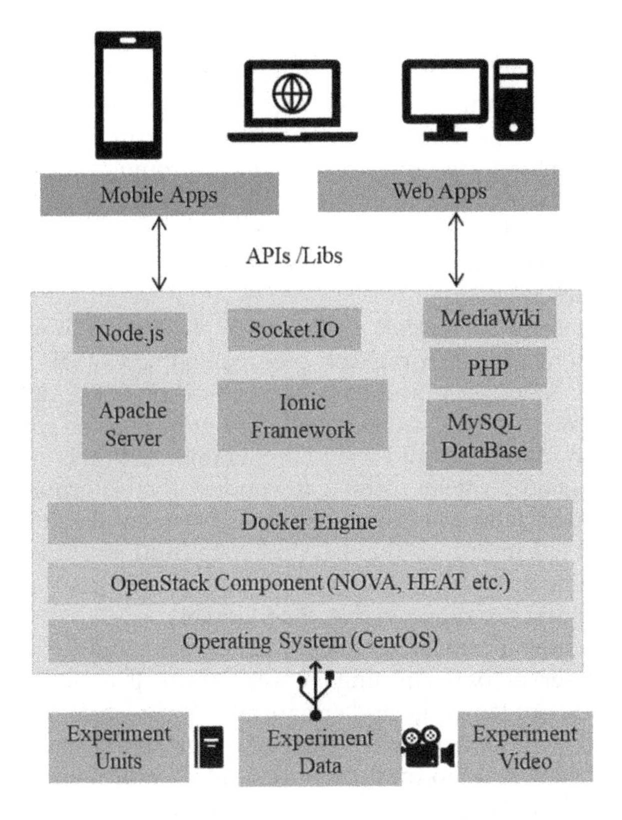

Figure 7.12 The software integration of the remote SVP experiment.

b. *Platform layer integration.* The platform layer application is directly built on the top of a novel assembled server engine scheme. It includes two server engines working together, i.e., Apache HTTP server engine and Node.js server engine. Based on the server-based mashup technology used in the framework, the Apache HTTP server engine is used to integrate the UI widgets with web content and the real-time experiment video. Meanwhile, the Node.js server engine handles the experiment data in real-time transmission. MediaWiki engine and Ionic framework also have been integrated into the RLaaS-Frame to support the different remote experiment apps. For the data management, MySQL is used for database management system. All of these RLaaS-Frame services are packaged in different Docker containers. OpenStack Nova computing component was implemented including the Docker containers. Combined with OpenStack Heat, the SVP essential services like MySQL and Apache can be integrated into OpenStack via the communication module.

Based on OpenStack, RLaaS-Frame can implement the Wiki platform and mobile-optimized remote experiment architecture with better cloud-computing resources. RLaaS-Frame can embrace the flexibility of OpenStack by using more powerful tools like compute, storage, networking, etc. So OpenStack promotes an environment of the more agile development of SVP remote experiment. More importantly, it will significantly improve the SVP remote experiment performance and make it better support functional features of the RLaaS-Frame.

c. *Integrate the LtoN module to the novel platform.* A novel real-time experiment data transmission protocol, named LtoN (LabVIEW to Node.js), is designed and developed based on Socket.IO and JSON. This new real-time experiment data transmission protocol also has been integrated into the RLaaS-Frame. With the new real-time transmission protocol, users can conduct the experiment, save the experiment data, and download the experiment data file. This new protocol has improved the real-time data communication capability of cloud-computing model. More details of this new data transmission protocol are illustrated in the following:
   - The new protocol includes client part running in browsers and server part running in web server. It is developed using JavaScript language and enhanced by the web socket.
   - In this new protocol, we defined our special communication instruction set to implement real-time experiment control commands and experiment data transmission.
   - In the new protocol, some brief instructions to control experiment progress are designed to improve the data transmission performance.

Based on the RLaaS-Frame, the new cloud-based remote SVP experiment can support users for learning key concepts in mechanical engineering.

7.5 Future works

Till now, the RLaaS model is successfully defined as a standardized cloud-computing model. A new cloud-based framework, named RLaaS-Frame, is designed and implemented based on the RLaaS model. Moreover, a new cloud-based remote SVP experiment application has been implemented by the new RLaaS-Frame. However, further research and development are still required to improve the RLaaS model, and enhance the stability and usability of RLaaS-Frame. More specially, issues that need improvement are as follows:

1. Various experiments for different disciplines need to be designed following the standardized RLaaS model. With a growing number of new requirements from users, more and more new remote experiments need to be designed and implemented based on the RLaaS-Frame in future.
2. Artificial intelligent technology (AIT) and big data technology will greatly improve remote laboratory development in such a way to enhance M-Learning and online learning. The usability of the new RLaaS-Frame and provide more powerful remote laboratory services to users, it can be used to support the cloud-based remote experiment for the various educational purpose.

The RLaaS-Frame is still version 1.0. So far, only a few sample pilot tests have been conducted. Some bugs in the software package have been fixed through users' feedback. More comprehensive testing and user survey still need to be done.

7.6 Conclusion

In this chapter, the RLaaS is defined as a new standardized cloud-computing model. Based on this standardized RLaaS model, a cloud-based framework, namely RLaaS-Frame, was successfully established for remote laboratory system rapid and flexible deployment and implementation. With this new RLaaS-Frame, the real-time data communication challenge discovered in the remote laboratory system deployment can be solved as well. As a pilot remote laboratory platform based on the RLaaS model, a Wiki-based remote laboratory platform and the mobile-optimized application framework have been integrated into the RLaaS-Frame successfully. A new cloud-based remote SVP experiment application has been implemented. Users can apply this cloud-based remote laboratory platform to learn engineering knowledge anywhere and anytime without plug-in issues. The future work will be to refine and improve this new RLaaS model. The RLaaS-Frame will significantly benefit online engineering including academic research and industrial applications.

References

1. C. Gravier, J. Fayolle, B. Bayard, M. Ates, and J. Lardon. State of the art about remote laboratories paradigms-foundations of ongoing mutations. *International Journal of Online Engineering (iJOE)*, 4(1), 2008, 1–9.
2. V. J. Harward, J. A. DelAlamo, S. R. Lerman, P. H. Bailey, J. Carpenter, K. DeLong, C. Felknor, J. Hardison, B. Harrison, I. Jabbur, P. D. Long, T. Mao, L. Naamani, J. Northridge, M. Schulz, D. Talavera, C. Varadharajan, S. Wang,

K. Yehia, R. Zbib, and D. Zych. The iLab shared architecture: A web services infrastructure to build communities of internet accessible laboratories. *Proceedings of the IEEE*, 96(6), 2008, 931–950.

3. P. Orduna, J. Irurzun, L. Rodriguez-Gil, J. Garcia-Zubia, F. Gazzola, and D. Lopez-de-Ipina. Adding new features to new and existing remote experiments through their integration in WebLab-Deusto. *International Journal of Online Engineering*, 7(S2), 2011, 33–39.
4. M. J. Callaghan, J. Harkin, T. M. McGinnity, and L. P. Maguire. Client-server architecture for remote experimentation for embedded systems. *International Journal of Online Engineering*, 2(4), 2006. ISSN:1861-2121.
5. J. García-Zubia, P. Orduña, D. López-de-Ipiña, and G. R. Alves. Addressing software impact in the design of remote laboratories. *IEEE Transactions on Industrial Electronics*, 56(12), 2009, 4757–4767.
6. M. Tawfik, D. Lowe, C. Salzmann, D. Gillet, E. Sancristobal, and M. Castro. Defining the critical factors in the architectural design of remote laboratories. *IEEE-RITA*, 10(4), 2015, 269–279.
7. L. Gomes, and S. Bosgoyan. Current trends in remote laboratories. *IEEE Transactions on Industrial Electronics*, 56(12), 2009, 4744–4756.
8. J. Rodriguez-Andina, L. Gomes, and S. Bogosyan. Current trends in industrial electronics education. *IEEE Transactions on Industrial Electronics*, 57(10), 2010, 3242–3244.
9. S. Seiler. Current trends in remote and virtual lab engineering. Where are we in 2013? *iJOE*, 9(6), 2013, 12–16.
10. M. Armbrust, A. Fox, R. Griffith, A. D. Joseph, R. Katz, A. Konwinski, G. Lee, D. Patterson, A. Rabkin, I. Stoica, and M. Zaharia. A view of cloud computing. *Communications of the ACM*, 53(4), 2010, 50–58.
11. P. Mell, and T. Grance. The NIST definition of cloud computing. *National Institute of Standards and Technology*, 53(6), 2009, 50.
12. A. T. Velte, T. J. Velte, R. C. Elsenpeter, and R. C. Elsenpeter. *"Cloud Computing Basics", Cloud Computing: A Practical Approach* (pp. 3–22). New York: McGraw-Hill, 2010.
13. N. Fernando, S. W. Loke, and W. Rahayu. Mobile cloud computing: A survey. *Future Generation Computer Systems*, 29(1), 2013, 84–106.
14. P. Jamshidi, A. Ahmad, and C. Pahl. Cloud migration research: A systematic review. *IEEE Transactions on Cloud Computing*, 1(2), 2013, 142–157.
15. N. K. Jangid. Real time cloud computing. In *Proceeding of the National Conference on Data Management and Security*, Amity University, Rajasthan, India, 2011.
16. K. Goldberg, and B. Kehoe, 2013. Cloud robotics and automation: A survey of related work. Berkeley: EECS Department, University of California, Tech. Rep. UCB/EECS-2013-5.
17. M. Tawfik, C. Salzmann, D. Gillet, D. Lowe, H. Saliah-Hassane, E. Sancristobal, and M. Castro. Laboratory as a Service (LaaS): A model for developing and implementing remote laboratories as modular components. In *Remote Engineering and Virtual Instrumentation (REV), 2014 11th International Conference on* (pp. 11–20). IEEE, Porto, Portugal, 26–28 February 2014.
18. L. Tobarra, S. Ros, R. Pastor, R. Hernandez, M. Castro, A. Al-Zoubi, M. Dmour, A. Robles-Gomez, A. Caminero, and J. Cano. Laboratories as a service integrated into learning management systems. In *Remote Engineering*

and Virtual Instrumentation (REV), 2016 13th International Conference on (pp. 103–108). IEEE, 2016.

19. S. Werner, A. Lauber, J. Becker, and E. Sax. Cloud-based remote virtual prototyping platform for embedded control applications: Cloud-based infrastructure for large-scale embedded hardware-related programming laboratories. In *Remote Engineering and Virtual Instrumentation (REV), 2016 13th International Conference on* (pp. 168–175). IEEE, February 2016.
20. D. Lowe. MOOLs: Massive open online laboratories: An analysis of scale and feasibility. In *Proceedings of IEEE International Conference REV Instruments,* February 2014, pp. 1–6.
21. C. Salzmann, D. Gillet, and Y. Piguet. MOOLs for MOOCs: A first edX scalable implementation. In *Proceedings of IEEE International Conference REV Instruments,* 2016, pp. 246–251.
22. N. Wang, J. Weng, X. Chen, G. Song, and H. Parsaei, 2015. Development of a remote shape memory alloy experiment for engineering education. *Engineering Education Letters,* 2, 1–20.
23. N. Wang, X. Chen, G. Song, and H. Parsaei. An experiment scheduler and federated authentication solution for remote laboratory access. *International Journal of Online Engineering (iJOE),* 11(3), 2015.
24. N. Wang, X. Chen, G. Song, and H. Parsaei. Using node-HTTP-Proxy for remote experiment data transmission traversing firewall. *International Journal of Online Engineering (iJOE),* 11(2), 2015, 60–67.
25. N. Wang, X. Chen, G. Song, and H. Parsaei. A novel real-time video transmission approach for remote laboratory development. *International Journal of Online Engineering (iJOE),* 11(1), 2015, 1–4.
26. N. Wang, X. Chen, Q. Lan, G. Song, H. Parsaei, and S. C. Ho. A novel Wiki-based remote laboratory platform for engineering education. *IEEE Transactions on Learning Technologies,* 10(3), 2017, 331–341.
27. N. Wang, X. Chen, G. Song, Q. Lan, and H. Parsaei. Design a new mobile optimized remote laboratory application architecture for M-learning. *IEEE Transactions on Industrial Electronics,* 64(3), 2017, 2382–2391.
28. N. Wang, X. Chen, G. Song, and H. Parsaei. Remote experiment development using an improved unified framework. *AACE Proceedings. E-Learn 2014--World Conference on E-Learning,* New Orleans, LA, United States, October 27–30, 2014.
29. T. Hughes-Croucher, and M. Wilson. *Node: Up and Running: Scalable Server-Side Code with JavaScript.* Sebastopol, CA: O'Reilly Media, 2012, ISBN: 978-1-4493-9858-3.
30. D. Herron. *Node Web Development* (2nd ed.), Birmingham, UK: Packt Publishing, 2013, ISBN: 184951514X.
31. Q. Hu, and E. Johnston. Using a Wiki-based course design to create a student-centered learning environment: Strategies and lessons. *Journal of Public Affairs Education,* 493–512, 2012.
32. R. Khanna, and M. Harlington. *"Anatomy of a Hybrid Mobile App" Getting Started with Ionic* (pp. 4–5), Birmingham, UK: Packt Publishing, 2016. ISBN: 978-1-78439-057-0.
33. N. Wang, G. Song, and X. Chen. Framework for rapid integration of offline experiments into remote laboratory. *International Journal of Online Engineering (iJOE),* 13(12), 2017, 192–205.

34. Q. Zhang, L. Cheng, and R. Boutaba. Cloud computing: State-of-the-art and research challenges. *Journal of Internet Services and Applications*, 1(1), 2010, 7–18.
35. J. E. Smith, and R. Nair. The architecture of virtual machines. *Computer*, 38(5), 2005, 32–38.
36. D. Merkel. Docker: Lightweight Linux containers for consistent development and deployment. *Linux Journal*, 2014(239), 2014, 2.
37. K. T. Seo, H. S. Hwang, I. Y. Moon, O. Y. Kwon, and B. J. Kim. Performance comparison analysis of Linux container and virtual machine for building cloud. *Advanced Science and Technology Letters*, 66(105–111), 2014, 2.
38. W. Felter, A. Ferreira, R. Rajamony, and J. Rubio. An updated performance comparison of virtual machines and Linux containers. In *Performance Analysis of Systems and Software (ISPASS), 2015 IEEE International Symposium on* (pp. 171–172). IEEE, Philadelphia, PA, March 2015.
39. C. Anderson. Docker [software engineering]. *IEEE Software*, 32(3), 2015, 102-c3.
40. C. Boettiger. An introduction to Docker for reproducible research. *ACM SIGOPS Operating Systems Review*, 49(1), 2015, 71–79.
41. Google Trend Chart. The trends of Docker and VM in worldwide. *Google Technical Trends Report*, February 2017. www.google.com/trends/explore?q=Docker,%2Fm%2F07yf2.
42. D. Bernstein. Containers and cloud: From lxc to docker to kubernetes. *IEEE Cloud Computing*, 1(3), 2014, 81–84.
43. F. Wuhib, R. Stadler, and H. Lindgren. Dynamic resource allocation with management objectives-Implementation for an OpenStack cloud. In *Network and service management (cnsm), 2012 8th International Conference and 2012 Workshop on Systems Virtualiztion Management (SVM)*, pp. 309–315, IEEE, October 2012.
44. S. A. Baset. Open source cloud technologies. In *Proceedings of the Third ACM Symposium on Cloud Computing*, p. 28, ACM, October 2012.

chapter eight

Augmented reality in engineering instruction

Mohamed Y. Ismail, Hamid R. Parsaei, and Konstantinos Kakosimos
Texas A&M University at Qatar

Contents

8.1 Introduction

Texas A&M University at Qatar is a branch campus of Texas A&M University in College Station, Texas, which opened operations in fall 2003 in Education City, Qatar. The Education City houses six international branches of American universities including, Texas A&M University, Virginia Common Wealth University, Cornell University, Carnegie Mellon University, Georgetown University, and Northwestern University. Each of these six branch campuses offers specialized degrees such as engineering, medicine, interior design, business and computer science, foreign services, communication, and journalism. Education City is an initiative of Qatar Foundation for Education, Science, and Community Development and has been co-founded chaired since its inception in 1997 under the leadership of H. H. Sheikha Moza bint Nasser.

Since 2003, Texas A&M University at Qatar has pioneered teaching and training efforts for the next generation of engineers in Qatar using state-of-the-art technology and benefited from skills and expertise of many experienced and seasoned faculty and technical staff. The programs have consistently championed in developing and adopting new teaching approached to further stimulate student interests and increase skills and experience necessary to become well-round engineers.

8.2 Teaching by adopting digital technologies

The precipitous increase in the popularity of modern technologies among students is increasingly challenging the status quo of the educational system. Higher education institutions need to evolve and reinvent many of the practices to appeal to their student populace. Pairing teaching and learning strategies with appropriate device and technology trends can greatly enhance the motivation to learn [1–3]. Ergonomics and human factors can play a critical role in analyzing the abilities and behaviors of student users and employing the gained knowledge in designing tools, products, and systems that are safer, acceptable, and more effective.

Generational research has already identified many of the characteristics of new student learners [4]. Although much of the work done focused on specific geographies, it is safe to say that many of the attributes are applicable to the Middle Eastern student population. This observation is particularly true in relation to technology-linked traits: Students are tech-savvy, accustomed to using social media sites and tools, dependent on search engines for information, comfortable exploring new technologies, and enjoy gaming more than any other generation [5]. Additionally, new student generations are good at multitasking, value collaboration, and often mix work and leisure times [5,6]. Optimizing the outcomes of technology implementations requires accounting for user characteristics.

Wearable technologies equipped with Augmented Reality (AR) capabilities offer a lot of promise for education and training especially for the applied science disciplines. Such technologies have features that greatly align with the expectations of young student. This is by no chance coincidental as many of the people involved in the design and development of new emerging technologies belong to the younger generation.

The purpose of this paper is to explore the development of visual support tools to enhance the instructional experience of chemical engineering students actively involved in laboratory experimentation activities while mitigating the challenges of laboratory work. Laboratories offer rich environments for instructional innovation and the impact of technology on learning could be profound. The approach discussed focuses on the utilization of wearable technologies to develop alternatives to paper-based instruction including video and AR enhanced instruction.

8.3 Motivation

Integrating technology in engineering education has been a primary objective at Texas A&M University at Qatar. Several steps were taken to enable the accomplishment of this objective including the buildup of educational technology infrastructure, establishment of professional development programs, and introduction of motivational strategies to encourage experimentation with innovative ideas that improve teaching and learning [7]. This study is the result of a technology competition that was sponsored in support of the motivational efforts carried out by the university. Although the study is specific to chemical engineering, the ideas presented can be easily extended to other engineering disciplines.

Integral to engineering instruction is educational laboratories that provide students with the hands-on experience necessary to supplement and clarify the scientific theory discussed in lectures. The practical nature of laboratory settings offers a rich ground for developing innovative solutions to enhance the instructional experience of students. The main drawback in such settings is the added risk involved. Technology has the potential to address both aspects.

Information about Standard Operating Procedures and safety policies as well as step-by-step instructions for performing the experiments are usually shared with students at different points during the semester. Information distribution is mainly done via paper documents with text and graphics content. During laboratory time, students are expected to bring the paper instructions and start conducting the experiment while continuously referencing the materials for procedures and directions. This protocol for disseminating information and operating in laboratory environments represents the standard followed by the majority of academic institutions. Neither the content format nor the delivery methods appeal to young students who have preferences that are different than past generations. The lack of appeal can lead to lack of motivation.

Updating existing protocols to take student expectations into consideration is therefore important to counteract such negative effects. Incorporating technology-enhanced methods that communicate information through visuals rather than graphics and text, support mobile Internet features, and provide game-style elements can improve the instructional experience while significantly increasing student interest and motivation [5,8,9].

8.4 Methodology

The approach pursued to achieve the objectives the study set out to accomplish focused on the utilization of wearable technologies with AR capabilities. The choice of AR is due to the ability of the technology to supplement real laboratory views with computer-generated messages that make the augmented experience richer, safer, and more informative. The wearable technology device of choice was the Google Glass. The glass includes features with strong appeal to the younger generations and supports capabilities for intelligent AR operation. Vuforia and Unity were the development frameworks utilized for building the AR functionality along with other application features. A major challenge of the proposed setup was the complexity of the development effort associated with building a functional AR prototype for the trials.

Execution of the proposed approach focused on the Thermo-Gravimetric Analysis procedure. Despite the limitation to one procedure, there are no technical reasons as to why the approach could not be duplicated with other experimental procedures practiced in chemical engineering laboratories.

8.5 Overview of augmented reality

AR is a promising innovative technology that blends and overlays live direct or indirect real-world physical environments with computer-generated virtual objects in real time [10]. According to the concept of reality–virtuality continuum, AR represents states in the Mixed-Reality region of the continuum that are closer to the pure reality limit [11,12]. In essence, the technology enhances one's perception of reality through computer-generated sensory cues such as sound, video, graphics, or haptic feedback to create a new mixed-reality environment that is richer and more informative than the primary natural environment [13].

Until recently, use of the AR technology was restricted to experimental settings in high-end research laboratories. Recent technological leaps, however, produced mobile AR systems that are extremely powerful and cost-effective and this opened the door for the development of new interactive devices with remarkable capabilities. Coupling powerful hardware platforms with new development tools that are considerably less complex for the average developer triggered the renewed interest in the technology.

The potential of AR in higher education is just beginning to emerge [14–16]. Applications of the technology in educational contexts are being extensively investigated by many researchers [17–19]. The impact of AR on teaching and learning is also being examined [20,21]. Results suggest that AR applications are effective at improving conceptualization and understanding while being noble motivational tools for students. One potential barrier to the widespread use of the technology in education is the time and technical skills required to develop AR tools and content [22].

8.6 Video-enhanced instruction

Initial efforts to enhance the instructional experience of students involved replacing standard paper documents and instructions with training and instructional videos recorded by the instructor or one of the laboratory managers using the Google Glass. Training videos aimed to demonstrate different aspects of lab operations, while the instructional videos aimed to provide step-by-step instructions for executing SOPs and safety procedures. The recording process was quite simple and only required wearing the glass while performing a task in the actual laboratory environment. The advantage of the videos recorded using Glass is the first-person views they provided. Wearing Glass to watch the videos provided students with a view of the activity being demonstrated from the point of view of the preparer. References to videos in the following sections imply wearing Glass to watch the recorded material.

For the purpose of this study, the impact of the proposed approach on the instructional experience of students was measured in terms of two factors: learning and training, and execution and support. In order to evaluate the effectiveness of the proposed approach on instructional experience as a whole, the impact on each of the individual factors had to be considered.

The learning and training factors assess how the instructional experience is impacted when replacing standard paper documents with training videos while being in lab settings. A student going through SOP training wears the glass inside the actual laboratory and activates the training video while facing relevant equipment. Seeing the actual equipment provides a direct feel of the procedure being demonstrated which could enhance student understanding.

The execution and support factor evaluates how the instructional experience is affected when typical paper instructions are replaced by video instructions to guide the execution of an SOP. A student needing to execute an SOP wears Glass and visually follows the video instructions specific to the SOP.

8.6.1 Impact assessment

Prior to the beginning of the study, engineering students were invited to participate in an AR experiment. Twenty students volunteered to participate but only thirteen completed all stages of the trials. Furthermore, 25% of the students reported no prior exposure to chemical engineering lab activities. The study structure involved splitting participants into two groups designated as A and B. The two groups had to execute a single SOP twice: Once utilizing Glass and once employing traditional methods without Glass. An unidentified SOP with ten distinct steps was selected for this purpose. The two trials were denoted by Glass-ON and Glass-OFF, respectively. Participants had to perform the two trials individually with members of group A having to execute the SOP following a Glass-OFF/Glass-ON sequence while members of group B having to follow the reverse sequence of Glass-ON–Glass-OFF. The two trials performed by a single student had to be separated by a minimum of 2 days. Before the beginning of a trial, the student is provided with training material relevant to the trial for brief review. The training material consisted of either paper documents for the Glass-OFF trials or training videos for the Glass-ON sessions. Once a trial starts, the student is provided with the proper instructional content, either paper or video, necessary for completing the SOP.

Prior to the start of trials, participating students attended a preparatory session which provided background information about the study as well as details about the envisioned setup. To further motivate them, the session employed video material about Google Glass, AR, and their potential applications in engineering education. At the conclusion of the orientation session, students were instructed to select two 15 min slots that are at least 2 days apart. In addition, students were surveyed to examine their initial expectations about video instructions and Google Glass.

Student surveys were used throughout the trials to assess the impact of the proposed approach on the instructional experience of students. Specific sections of the surveys focused on how students perceived training videos shared with them before trials on learning and training. Other sections examined student opinions about Glass and the impact of instructional videos on the execution and support of SOPs.

8.6.2 Learning and training

A Likert-scale survey consisting of three questions specific to training videos was administered to students after the orientation session in order to establish a baseline for comparing the post-training survey results. At the end of each trial, students had to retake the survey to measure changes in their sentiments as a result of the video training. The survey results are provided in Table 8.1. The results are filtered to reflect the "Agree" and "Strongly Agree" values only.

Responses to the first question in the pre-training survey indicate that 100% of the students prefer training videos over the standard paper documentation. The positive sentiment continued to hold at 100% after completing the video training. This result was expected as new generations of students are known to prefer visual communications over text and graphics. Analysis of the pre-training student responses to the second

Table 8.1 Pre- and post-trials survey results for learning and training

Question	Pre-training (%)	Post-training (%)
Q1. I prefer recorded material over written documents for this type of training	100	100
Q2. Training videos are useful and easy to follow	100	85
Q3. Paper documents are useful and easy to follow	30	60

and third questions provides further evidence that students favor training videos over traditional paper documents. Whereas only 30% of the students believed that paper documents are useful and easy to follow, 100% of the participants believed the same about training videos. Post-training, the percentages for the second and third questions remained strongly in favor of training videos despite changes in the percentages. Responses to the second question revealed positive sentiments in 85% of the cases, while the results reported for the third question showed agreement by 60% of the students surveyed. The drop in the post-training percentage for the second question and the corresponding rise in the post-training percentage for the third questions are both attributed to negative Glass-ON experiences resulting from defects in the prepared videos. Overall, results came in line with expectations.

8.6.3 Execution and support

Using Glass as visual aid for the execution and support of SOPs and safety instructions proved unsatisfactory. Analysis of data gathered from student surveys administered after each trial did not reveal any significant variation between the two trial sequences. In essence, the students did not feel that the use of Glass for execution and support was making a difference. This is contrary to the initial expectations enforced by the pre-trial opinion of 80% of the students who expected the Glass-ON experience for executing an SOP to be far more positive than the traditional paper-based process. The perceived ineffectiveness of Glass can be attributed to a combination of factors including application instabilities, usability issues, and the lack of skills necessary for proper and smooth operation of Glass.

Whereas the use of Glass for mainstream learning and training would be constrained to few sessions, a more general use for execution and support would be routine. As such, usability aspects become more relevant to the assessment efforts. Questions about the usability of Glass were incorporated in the pre- and post-trial surveys. The questions targeted to measure changes in student expectations in relation to the advantages and disadvantages of Glass. The major advantage anticipated by 90% of the students surveyed was the hands-free operation. This is in contrast to the 10% who predicted the same about the visual support capabilities. In terms of disadvantages, 30% of participants expected the small size of the head display to be problematic while one student expressed concern about the device's fit and the need for continuous readjustment. Analysis of the post-trial survey data revealed significant changes in the results pertaining to the disadvantages of the device. The need to continuously readjust the fit became the biggest disadvantage as expressed by 70% of the students. The percentage of students who thought the size of the head display is problematic slightly dropped to 25%. Another inconvenience that was reported in the post-trial surveys by 15% of the students is the need to frequently touch Glass in order to prevent it from going into standby mode.

The impact of Glass on execution and support has so far been investigated in terms of subjective measures. In order to quantify the impact, data points connected to the number of successful SOP steps completed and time to complete the entire SOP were collected and analyzed. While the majority of students successfully completed all the steps of the SOP in both of their trials, two students suffered the inconvenience of entering Glass' standby mode and eventually missed one step as they attempted to reactivate Glass and resume the SOP. The average time required to complete the SOP favored the traditional paper approach by 2 min. Students had difficulty in executing the SOP while navigating Glass due to the lack of training on device controls. The delay issue, however, could easily be resolved through proper training and sufficient exposure to Glass.

Contrary to initial expectations, the use of Glass did not provide any clear advantage over traditional methods in terms of the execution and support function. A deep dive into the reasons behind the negative assessments points to training and implementation issues rather than methodology or technology problems. This is comforting as both issues can be mitigated.

8.7 AR-Enhanced instruction

Thus far, the discussion focused on enhancing the instructional experience through the use of training videos recorded via the Google Glass. The videos, however, lacked any interactivity with the actual laboratory environment. A more engaging approach that can significantly enrich student experiences while mitigating some of the risks involves employing AR technology.

The direction pursued encompassed tagging chemicals, substances, and instrumentation with QR codes, and utilizing AR technology to blend the QR feed information and procedural instructions with the actual laboratory environment. An individual wearing the glass starts by selecting a particular procedure to execute. Procedural summary appears as a list in an inset that overlays the laboratory view. The wearer then receives instruction to take specific actions. Every time an action is completed, and instructions for the next action are provided. Completion in this case is defined in terms of detecting the right QR code associated with the proper substance or the right instrumentation. At every step in the process, real-time views of the actual lab environment are overlaid by relevant information. Figure 8.1 shows still captures from Glass views recorded during one of the demonstrations for the tool where the task was to select the proper sample for a thermal analysis procedure.

As shown in the figure, the activity started with a procedural summary identifying the need for sample S05. As the user attempted to locate the sample, the QR code for a cabinet that the user visited was scanned and the nature of substances inside it were identified. The user then went on to examine the different samples inside the cabinet where he was

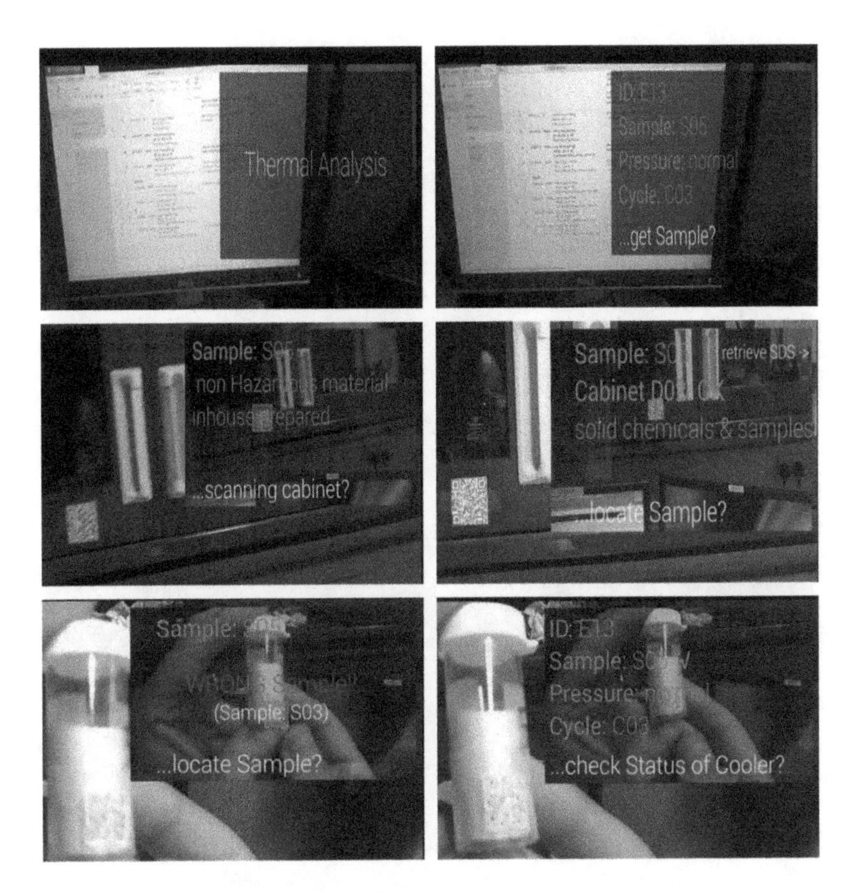

Figure 8.1 Illustration of the use of AR for sample selection.

receiving instant feedback about the correctness of the sample picked. The wearer is then instructed to go to the next step of the procedure which is checking the status of the cooler. Figure 8.2 provides captures from the same demonstration for the cooler and cylinder checks of the procedure.

In the first capture, the wearer received the task completion check. The wearer was then instructed to investigate the various cylinders for proper pressure. As the user began to examine the cylinders, information about their contents and proper settings was displayed. The feedback provided was then utilized by the wearer to properly set the gauges.

Due to time limitations of the competition, no formal assessment was conducted to judge this effort. Although the results presented are only intended for proof of concept, the general approach has the potential to significantly improve the instructional experience for students. Furthermore, using AR to guide actions has the potential to reduce human errors associated with experimental procedures and can lead to safer practices.

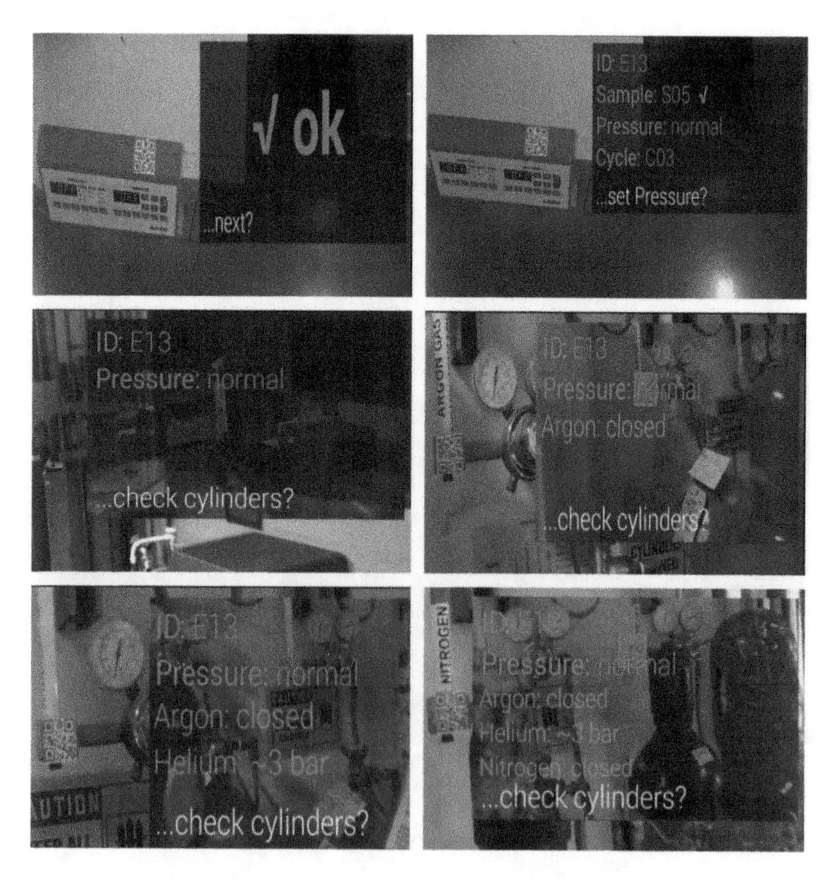

Figure 8.2 Illustration of the use of AR for guiding lab actions.

8.8 Conclusion

Lab activities are an integral part of engineering education. Enhancing the educational experience associated with laboratory work to account for changing student expectations necessitates investigating alternative approaches to the traditional methods of instruction. The rising popularity of wearable devices among young students provides educators with new possibilities to improve upon the traditional content and to offer a whole new level of interactive learning experience. This paper presented information about a pilot study that explored alternative modes of laboratory instruction. The main purpose was to provide students with an enhanced instructional experience to keep them engaged and motivated. The study started by examining the impact of replacing traditional paper-based documents with videos instructions. Results did not align with initial expectation due to implementation and training issues. Additional extensions to the study involved investigating AR-enhanced instruction. The potential of AR to produce real-world, safer, and more engaging experience is clear; however, more development work is required.

References

1. Jonassen, D.H., Carr, C., & Yueh, H.P. Computers as mindtools for engaging learners in critical thinking. *TechTrends*, 43(2), 24–32, (1998).
2. Chu, H.C., Hwang, G.J., & Tsai, C.C. A knowledge engineering approach to developing mindtools for context-aware ubiquitous learning. *Computers & Education*, 54(1), 289–297, (2010).
3. Hwang, G.J., Tsai, C.C., Chu, H.C., Kinshuk, K., & Chen, C.Y. A context-aware ubiquitous learning approach to conducting scientific inquiry activities in a science park. *Australasian Journal of Educational Technology*, 28(5), 931–947, (2012).
4. Prensky, M. Digital natives, digital immigrants. *On the Horizon*, 9(5), 1–6, (2001).
5. Black, A., & Gen, Y Who they are and how they learn. *Educational Horizons*, 88(2), 92–101, (2010).
6. Maürtin-Cairncross, A. A glimpse of generation-Y in higher education: Some implications for teaching and learning environments. *South African Journal of Higher Education*, 28(2), 564–583, (2014).
7. Alaeddine, N.I., Parsaei, H.R., Kakosimos, K., Guo, B., & Mansoor, B. Teaching innovation with technology to accelerate engineering students' learning. *Paper presented at 2015 ASEE Annual Conference and Exposition*, Seattle, Washington. 10.18260/p.24815, (2015).
8. Metros, S.E. The educator's role in preparing visually literate learners. *Theory into Practice*, 47(2), 102–109, (2008).
9. Gonzalez, J., Pomares, H., Damas, M., Garcia-Sanchez, P., Rodriguez-Alvarez, M., & Palomares, J.M. The use of video-gaming devices as a motivation for learning embedded systems programming. *IEEE Transactions on Education*, 56(2), 199–207, (2013).

10. Azuma, R. A survey of augmented reality. *Presence: Teleoperators and Virtual Environments*, 6(4), 355–385, (1997).
11. Milgram, P., & Kishino, F. Taxonomy of mixed reality visual displays. *IEICE Transactions on Information Systems*, E77-D(12), 1321–1329, (1994).
12. Milgram, P., Takemura, H., Utsumi, A., & Kishino, F. Augmented reality: A class of displays on the reality-virtuality continuum. *Telemanipulator and Telepresence Technologies*, 2351, (1994).
13. Graham, M., Zook, M., & Boulton, A. Augmented reality in urban places: Contested content and the duplicity of code. *Transactions of the Institute of British Geographers*, 38(3), 464–479, (2013).
14. Yuen, S., Yaoyuneyong, G., & Johnson, E. Augmented reality: An overview and five directions for Ar in education. *Journal of Educational Technology Development and Exchange*, 4(1), 119–140, (2011).
15. Bacca, J., Baldiris, S., Fabregat, R., & Graf, S. Kinshuk: Augmented reality trends in education: A systematic review of research and applications. *Journal of Educational Technology & Society*, 17(4), 133–149, (2014).
16. Antonioli, M., Blake, C., & Sparks, K. Augmented reality applications in education. *Journal of Technology Studies*, 40(2), 96–107, (2014).
17. Shirazi, A., & Behzadan, A.H. Content delivery using augmented reality to enhance students' performance in a building design and assembly project. *Advances in Engineering Education*, 4(3), 1–24, (2015).
18. Mejías Borrero, A.M., & Andújar Márquez, J.A. A pilot study of the effectiveness of augmented reality to enhance the use of remote labs in electrical engineering education. *Journal of Science Education & Technology*, 21(5), 540–557, (2012).
19. Odeh, S., Abu Shanab, S., & Anabtawi, M. Augmented reality Internet labs versus its traditional and virtual equivalence. *International Journal of Emerging Technologies in Learning*, 10(3), 4–9, (2015).
20. Ibanez, M.B., Di Serio, A., Villaran, D., & Kloos, C.D. Experimenting with electromagnetism using augmented reality: Impact on flow student experience and educational effectiveness. *Computers & Education*, 71, 1–13, (2014).
21. Gutiérrez, J.M., & Meneses Fernández, M.D. Applying augmented reality in engineering education to improve academic performance & student motivation. *International Journal of Engineering Education*, 30(3), 625–635, (2014).
22. Billinghurst, M., & Duenser, A. Augmented reality in the classroom. *Computer*, 45, 56–63, (2012).

chapter nine

Challenges of engineering program and student learning assessment

Osman Taylan and Abdullah Bafail
King Abdulaziz University

Contents

9.1 Introduction

Learning and its quality are assessed as an act of assigning a qualitative or quantitative merit to the level it has been achieved by a student. Evaluation of the student learning level is one of the most important and difficult tasks in the whole learning and teaching processes. Assessment is a purposeful, systematic, and ongoing collection of information as evidence for the use in making judgments about learning level of a student. Two parts of the assessment system requires the attention of academics exist: one is the selection of the assessment tools and the other is the selection of the grade assignment method. Therefore, both qualitative and quantitative information are valuable forms of evidence for assessing the student outcomes. The quantitative information consists of numerical data; hence, the learning performance of a student is determined by the tests or questionnaires. Alternatively, qualitative data, however, can be "richer" than quantitative information, because they provide more and extensive variety of information related to a particular learning goal. Many faculty members usually use numerical scores for their teaching evaluation to make overall judgments of student's owning performance. A common misconception is that qualitative assessments are not as reliable, valid, or objective as quantitative ones. There are well-designed and statistically reliable means of interpreting and analyzing qualitative data for learning assessment.

For student learning assessment, multiple approaches such as quantitative and qualitative methods can be used together. On the other hand, benchmarking is now very common in assessment plans. Originally, benchmarking was a term used in the production field to define a set of external standards against which an organization could measure itself. In higher educational settings, a university might use benchmarking techniques to define its comparison group—its peer institutions—and compare its outcomes to their competitors. The benchmarking could be based on retention rates, 5-year graduation rates, admissions yield data, and employment and graduate school placement rates. The benefit of inter-institutional comparison is that it can flag areas those carrying problem and investigate the causes of results that differ from the norms.

9.2 A Global engineering program and student teaching-learning assessment

Student outcomes are called as program outcomes in the previous Accreditation Board for Engineering and Technology (ABET, 2016) applications. They are related to learning objectives and indicate what students have to learn and what they are able to do or perform at the end of a learning process, even during graduation. Reich et al. (2014) state that transfer of knowledge from teachers to students is just one part of learning, while students' experiences and interaction with society, and their competency are the other important learning components. Freeman and Dobbins (2013) show that measuring and assessing students' learning are one of the most challenging tasks in the whole education and teaching process. According to ABET applications, students' learning outcomes (Hargreaves, 2007) clearly reflect what is expected from an engineering program because the course learning objectives and even the subjects learning objectives planned by instructors in accordance with the whole engineering program objectives. Hence, student outcomes and course learning objectives are designed in a way to complement each other and direct students as both are based on program objectives. Implementation of student objectives at the course level carries much importance despite their reflection at the institute or university level (European Commission, 2011). Student outcomes must be observable, achievable, and measurable so that the student learning level can be assessed. However, actual measurement of student outcomes is a challenging task and requires continuous assessment methods, tools, and professional judgment approaches from all program constituents. In this regard, different direct and indirect assessment tools can be employed to assess student outcomes for engineering education practices. Rogers (2003), however, stated that inconsistent use of assessment tools often creates disparity among instructors engaged in the assessment processes in addition to other difficulties which hinder the assessment of student learning process. Northwood (2013) and Tamburri (2013) discussed the trend in which student outcomes are regarded as one of the major quality assurance tools to measure the learning-level objectives. Polikoff and Porter (2014) claimed that quality of any education system can be evaluated from the level of skills and competencies that students develop within the classroom and exhibit it in their professional life. Although several learning assessment methods are available (Brookhart, 2013), Liu et al. (2014) identify various challenges affecting their usefulness. They emphasized on the development of effective qualitative methods which not only assess the achievement of students but also realistically measure the usefulness of a program. Grez and Valcke (2013) present an innovative technique to assess the oral presentations of students during research projects that integrate their higher order cognitive skills with scientific communication. Dadach (2013) and Lopes et al. (1997) proposed qualitative and quantitative methods to assess active learning of students and found many fuzzy connotations on them; however, Ma and Zhou (2000) assessed student learning outcomes through well-defined integration of fuzzy logic. Wang et al. (2007) conducted a similar study in which the fuzzy set theory was used for the evaluation and grading of students. Results of this study indicated the effectiveness of fuzzy systems in resolving the complex and ambiguous human reasoning due to their simultaneous handling of qualitative and quantitative data. Meanwhile, Taylan and Karagozoglu (2009) developed a flexible neuro-fuzzy system to better predict the learning performance of students. Tian and Lowe (2013) revealed that cultural differences create several emotional and psychological challenges among students during the initial period of their courses. Spelt et al. (2014) developed, implemented, and evaluated a research method to improve learning in the interdisciplinary field of food industry with an objective to determine the effectiveness of this method in developing multidimensional learning among students.

Both direct and indirect methods other than qualitative and quantitative techniques are commonly used for evaluation purposes. However, the data used for this purpose are mostly collected using diversified assessment methods. Thus, in the absence of any unified assessment technique, these varied data inputs emphasized the need of flexible and integrated methods of evaluation. Terminologies such as outcome, goal, objective, student performance criteria (Baird, 2013), and standards can be understood in either similar or different ways. As a result, faculty members have to spend much time in understanding and comprehending the use of these terminologies, which not only create demotivation among them, but also contribute toward incorrect assessments (Rogers, 2003). The aim of this study was to standardize the learning quality of students by bringing improvement in the evaluation method of student outcomes. An integrated integer-programming approach was developed to assess the learning level of students and their understanding because it is evident that students' real-life behavior directly reflects the quality of program from which they are graduated. However, an accurate assessment of an engineering program is one of the difficult tasks in the whole process of learning and teaching process. An evidence-based approach is required to assess students' level of learning through specific student outcomes. Similarly, grades cannot be taken as the only basis for student assessment because faculty members for the same course may vary in their teaching quality, content selection, and delivery methods (Polikoff and Porter, 2014). Another reason of the reduced dependency on grades is disparity within faculty members toward grading policy and assessment criterion. Numerical scores and letter grades mostly show the relative position of students among their classmates, but are unable to show the results of authentic assessment about the clarity of concepts, understanding of topics, and their ability to apply knowledge in real life (Liu et al., 2014). The scope of this study included the assessment of student's learning levels for which the numerical scores were directly related to students' learning objective. Therefore, the collected data were subsequently be utilized to improve the program. In this study, the program outcomes were considered to be the student outcomes which were already identified by ABET with letters aligned from a to k.

9.2.1 Assessment of engineering students' learning quality

An engineering education assessment system must be as free from irrelevant errors as possible and have the following characteristics:

- relevant,
- reliable,
- recognizable and
- realistic

A relevant assessment system is based on the validity of the assessment method. In a relevant system, student's learning must be an accurate reflection of the skill or competency tested. However, students usually complain that the tests are not related to the content of the course or what is/are presented in the lectures. The assessment tools must be related to the course content, and directly derived from the course objectives. However, the format for evaluation can take many forms: in-class exam, out of class homework, long projects, and in-class exercises done by group work, etc. The performance of a relevant evaluation should well predict the performance of other closely related skills. Evaluation activities chosen should really measure the skills and knowledge which we intend to measure.

A reliable assessment system considers that a student's grade should not hang on only a single performance criterion or on mood of the instructor making the judgment. Several tools must be used in a reliable system. No system is perfectly reliable and will exactly produce the same evaluation performance each time. But the goal must eliminate as many sources of error as possible. The errors in reliably evaluating a student performance are usually due to the lack of sufficient information; in this respect, using one format may give rise to a source of bias and lower the reliability of assessment.

A recognizable assessment system means that students should be aware of how they will be evaluated and their class activities should prepare them for those evaluations. Exams and tests should not be a game of guess what I'm going to ask you? An instructor must choose evaluation types which are clearly related to the content and daily activities of the course. The test should be the best statement of the course expectations (course learning objectives) and should mirror the teaching level and performance of the teaching system. A realistic system is that the amount of information obtained is balanced by the amount of work required. A student usually takes four to five courses along with our course. The instructor also teaches two courses and carries out a lot of administrative duties which means both the teaching and learning sides are busy. Hence, what is a realistic system? Some claims that several smaller assignments tend to be more valuable than one large assignment. If a large assignment is called for, then spreading it out across the semester and requiring components to be periodically handled are good techniques from a learning and administration standpoint.

On the other hand, in a structured educational system (Hayward, 2015), learning is usually considered as a two-step process, involving the treatment and processing of information. Extensive use of Bloom's Taxonomy (1984) in educational objectives was proven to be very effective in the development of students' learning level. In Figure 9.1, the levels of learning in an educational system are illustrated. The three main learning domains of Bloom's Taxonomy (1984) are cognitive, affective, and psychomotor which are generally described as content, values, and soft and technical skills, respectively. They are associated with various levels of learning and are compiled from multiple sources.

For instance, the cognitive learning is related to creative thinking, intellectual capabilities (Bloom, 1956), and competency of students. The linguistic tools used for the measurement of this learning level are knowledge, comprehension, application, analysis, synthesis, and evaluation. Students can demonstrate their cognitive learning level by knowledge

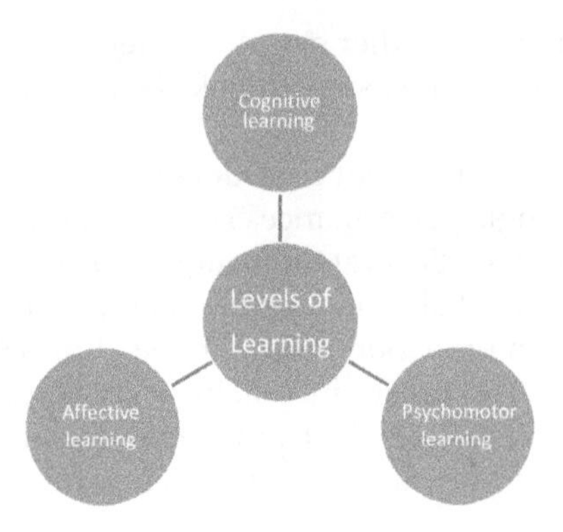

Figure 9.1 Levels of learning in an engineering educational system.

recall and the intellectual skills earned. These skills and competencies are comprehending information, organizing ideas, analyzing and synthesizing data, applying knowledge, choosing among alternatives in problem-solving, and evaluating ideas or actions. The affective domain is related to attitudes and feelings of students. The attitudes of students before, during, and after this learning process are described as they receive knowledge, respond, and value it (Krathwohl, 2002). Students demonstrate the affective learning by their behaviors indicating attitudes of awareness, interest, attention, concern, and responsibility. This learning level is the ability to listen and respond in interactions with others, and the ability to demonstrate those attitudinal characteristics or values which are appropriate to the test situation and the field of study. These skills can be called as earned soft skills. Similarly, Dominguez et al. (2014) described how instructors of engineering students use online peer-review methods to endorse their critical thinking and communication skills. Meyer et al. (2014) developed a methodology for students in order for them to identify prominent issues of any problem and then apply their knowledge for its analysis and immediate solution. Psychomotor learning is demonstrated by physical skills such as coordination, dexterity, manipulation, grace, strength, and speed. It also covers the actions which demonstrate competencies and skills such as the use of precision instruments, tools, or actions that are the evidence of skills such as the use of body in workplace. Similarly, in Figure 9.2, the level of learning, its outcome elements, and the depth of understanding are displayed. For instance, the students who achieve only knowledge and comprehension level indicate that they are novice and still at the introduction level of learning. Students who have knowledge level of learning can remember previously learned materials which may involve the recall of a wide range of material from specific facts to complete theories, but all that is required is to bring in mind

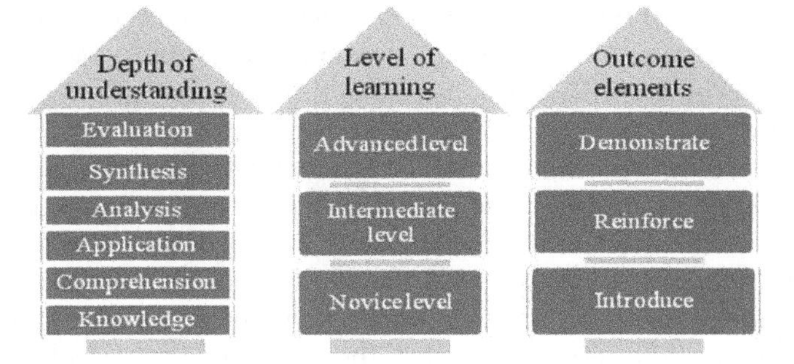

Figure 9.2 The house of cognitive learning.

of the appropriate information. The knowledge level that comes under the domain of cognitive skills alerts students about different dimensions of a problem. Similarly, they learn about different components of a problem in the analysis phase while the synthesis phase guides them to integrate different components. As a continuation of this level, comprehension level learning is the ability to grasp the meaning of materials which may be shown by translating material from one form to another (for instance words to numbers), by interpreting material (explaining or summarizing), and by estimating future trends (predicting consequences or effects). Hence, this goes one step beyond simply remembering of material and represents the lowest level of understanding. In the comprehension level of learning, students also try to summarize scattered pieces of information and find the emerging trends (Anderson and Krathwohl, 2001). The next level of application equips students with necessary skills to learn how to apply what they have learned (Black, 2015). This is called as learning skill level 1 in Bloom's Taxonomy.

Learning Skill level 2 in Bloom's Taxonomy (1984) stands for application and analyses of engineering problems. This is an intermediate level and focuses on the application of knowledge and analysis of findings. An engineering student at the application level of learning has the ability to use the learned material in new, concrete situations. He/she might include the application of rules, methods, concepts, principles, laws, and theories which require a higher level of understanding than those under comprehension. On the other hand, students who achieve the application level of learning have ability to disintegrate the material into its component parts so that its organizational structure can be understood. They may include the identification of parts, analysis of the relationship between parts, and recognition of the organizational principles involved. They can also represent a higher level than comprehension and application level of learning because they require an understanding both of the content and the structural form of the material.

Skill level 3 of Bloom's Taxonomy (1984) is the synthesis and evaluation of information by engineering students. At synthesis level of learning, students have ability to put parts together to form a new whole, which also involve the production of a unique communication, a plan of operations (research proposal), or a set of abstract relations (scheme for classifying information) and some more details. They can illustrate creative behaviors with major emphasis on the formulation of new patterns or structure. Students who achieved the evaluation level of learning have ability to judge the value of material for a given purpose based on definite criteria which contains elements of all the other categories, and judgments based on clearly defined assessment criteria. Assessing engineering program student-outcomes are unlike assessing student-outcomes at class level learning (Palermo, 2011). In class level assessment, faculty members should carefully determine the course objectives, course contents, assessment tools, and grading policy. In the evaluation phase, students can develop their critical thinking skills to see the outcome of their planning.

9.2.2 *Quantitative approaches for engineering program assessment*

In the higher education system, student grades are important, but when assessment is carried out at the program level, the weightage of individual course grades become less important because courses have equal weight in the grade average point (GPA) calculation. However, this cannot undermine the importance of classroom assessments, the questions in each and every direct assessment as the student outcomes eventually reflect and align in program outcomes. Thus, different direct or indirect assessment methods are used to measure students' learning level and skills during and after graduation. Although direct assessment is considered an easy way by which students are directly examined for the required competencies and skills (Ball et al., 2012), many activities are difficult to be directly measured, so indirect assessment methods are employed. However, both these methods have their own limitations that lead to the use of multiple methods with multiple inputs (Klenowski et al., 2006). Direct assessment tools consist of simulations, behavioral observations, performance appraisals (Grissom and Loeb, 2015), and oral and written exams such as final exam, major and minor exams, term project, and active learning sessions. Examples of indirect exams are exit and entry surveys, interviews, archival data, focus groups, and written surveys. Figure 9.3 shows the relation of assessment tools in an engineering program to establish a strategy. The assessment of a program starts with the questions asked in exams. The instructor must prepare the questions for each course to support the sub-criteria establishing the key performance indicators (KPIs) as each course supports one or two KPI and each KPI supports

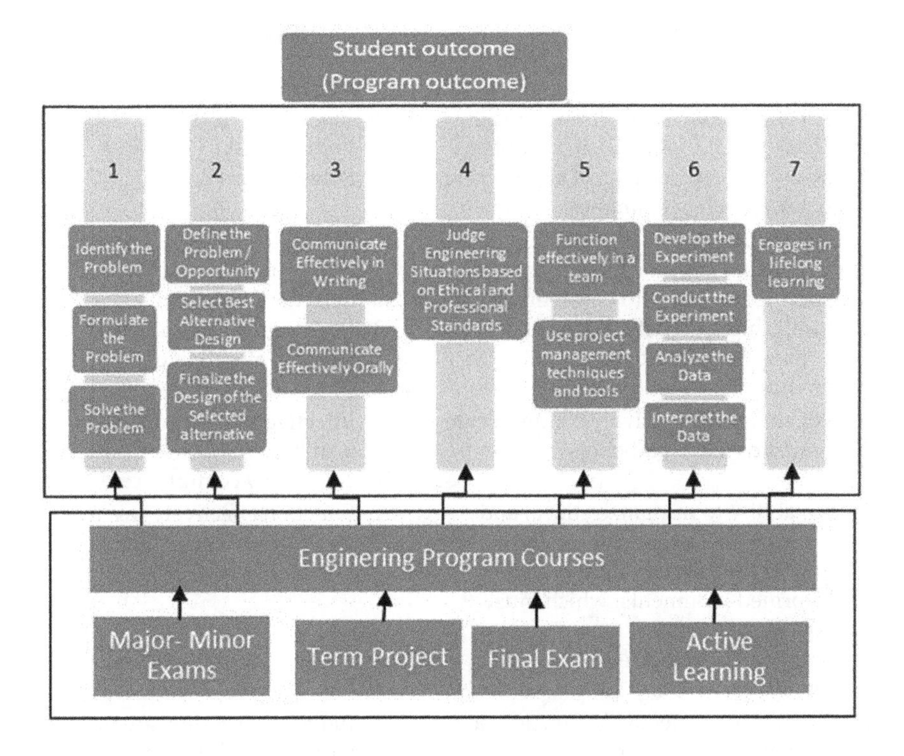

Figure 9.3 The student learning-level assessment strategy.

the student outcomes. There is a hierarchical relation between the exam questions, engineering courses, sub-criteria, KPIs, and student outcomes. As seen in Table 9.1, the KPIs and the sub-criteria used for the assessment of student learning level are presented. In Figure 9.3, the numbers "1–7" refer to the KPIs presented. In Table 9.2, industrial engineering courses and student outcomes are matched. This shows the courses supporting the student outcomes. Twenty-two courses are used for the assessment of student outcomes.

Hence, the assessment of an engineering program is a multicriteria decision-making problem. A multicriteria decision-making (MCDM) approach can support an engineering program to make decision for its quality level and student's learning level. The student learning-level problem is identified by KPIs and the qualitative sub-criteria are determined for further analysis, and fuzzy logic can be employed for the solution of this problem. The MCDM approach is an evaluation and rewarding method to consider not only the cost but also the other important criteria and sub-criteria of student learning level. The key of MCDM approach is that the decision process is based on the attributes called as a to k in old ABET assessment approach which is "1–7" in the new integrated ABET

Table 9.1 The KPIs and sub-criteria for the assessment of learning level

Key performance indicators	Sub-criteria for learning level
1. An ability to identify, formulate, and solve complex engineering problems by applying principles of engineering, science, and mathematics	1.1. Identify the problem 1.2. Formulate the problem 1.3. Solve the problem
2. An ability to apply engineering design to produce solutions that meet specified needs with consideration of public health, safety, and welfare, as well as global, cultural, social, environmental, and economic factors	2.1. Define the problem/ opportunity 2.2. Select best alternative design 2.3. Finalize the design of the selected alternative
3. An ability to effectively communicate with a range of audiences	3.1. Effectively communicate in writing 3.2. Effectively communicate orally
4. An ability to recognize ethical and professional responsibilities in engineering situations and make informed judgments, which must consider the impact of engineering solutions in global, economic, environmental, and societal contexts	4.1. Function effectively in a team 4.2. Use project management techniques and tools
5. An ability to effectively function on a team whose members together provide leadership, create a collaborative and inclusive environment, establish goals, plan tasks, and meet objectives	5.1. Judge engineering situations based on ethical and professional standards
6. An ability to develop and conduct appropriate experimentation, analyze and interpret data, and use engineering judgment to draw conclusions	6.1. Develop the experiment 6.2. Conduct the experiment 6.3. Analyze the data 6.4. Interpret the data
7. An ability to acquire and apply new knowledge as needed, using appropriate learning strategies	7.1. Engages in lifelong learning

assessment system. In Figure 9.3, the assessment criteria and sub-criteria employed to determine student learning level are presented. The goal was to determine the distribution of learning level, and its quality to avoid the high costs that might occur to the nation due to the low learning level. Hence, a set of criteria was determined by ABET to determine the student learning performance essentially to ensure that the quality level of the educational system is thoroughly assessed.

Fuzzy decision tree approach, fuzzy AHP, and fuzzy TOPSIS method can be used to compare the learning level of students and the effectiveness

Table 9.2 Industrial engineering courses and student outcomes matching matrix

Industrial engineering program core courses	Student outcomes: graduates must have these skills and competencies (KPIs)						
	1	2	3	4	5	6	7
IE 255	✓						✓
IE 256				✓	✓		
IE 311	✓						✓
IE 321	✓						✓
IE 322	✓	✓					
IE 323		✓					✓
IE 331	✓					✓	
IE 332	✓					✓	
IE 341	✓					✓	
IE 342		✓				✓	
IE351			✓		✓		
IE 352		✓			✓		
IE 390							
IE 395			✓	✓			
IE 411	✓						✓
IE 422	✓	✓				✓	
IE 431		✓				✓	
IE 432		✓				✓	
IE 441			✓	✓	✓		
IE 451	✓				✓		✓
IE 453		✓	✓	✓			
IE 499	✓	✓	✓	✓	✓	✓	✓

of an engineering program. Hence, the findings of assessment can be used to compare the performance of engineering programs, make recommendations for enhancing them, and select the best one. For instance, fuzzy decision tree assumes that all domain attributes or linguistic variables have predefined by fuzzy terms using fuzzy restrictions. An integrated fuzzy AHP and TOPSIS method can be used to improve the quality of decisions made for ranking alternatives scenarios and engineering programs. Taylan et al. (2014) extended the TOPSIS method to fuzzy group decision-making situations by using an exact Euclidean distance between any two fuzzy numbers, stating that fuzzy TOPSIS method has been widely used due to its rationality, logicality, and computational simplicity. Deng et al. (2000) used the TOPSIS approach to compare company performances and financial ratio performance within a specific industry. Fuzzy AHP method can be used to create favorable weights for fuzzy linguistic criteria of KPIs of educational system. Fuzzy TOPSIS can be employed to order the performance of engineering programs for their superiority. Many MCDM methods are defined in the literature. However, under certain conditions, deterministic approaches are used to model and assess the engineering programs, although the judgments of almost all instructors are vague, imprecise, and unsuitable to estimate the preferences with exact numerical values. As a more realistic approach, fuzzy linguistic assessment approach may be employed instead of these deterministic quantitative approaches to assess the engineering programs and the student learning level as a whole. Different weights might be considered for the courses during assessment process. The disadvantage of the current assessment approaches is that all courses have equal weight although the content of courses and their contribution to obtain technical and soft skills are different. Therefore, a reassessment tool by means of fuzzy linguistic variables can be employed because fuzzy set theory aids in measuring the ambiguity of linguistic concepts that were associated with instructor's subjective judgments. Moreover, when a group of instructors evaluate and select the weights, considering different scenarios using different fuzzy settings and fuzzy decision-tree approach can be used in the determination of the sub-criteria and KPIs' weight.

9.2.3 *Quantitative assessment of student learning level*

Accreditation Board for Engineering and Technology (ABET, 2016) determines the KPIs for the assessment of engineering program in that the ability of students to apply knowledge of math, sciences, and engineering is defined as a, and their skill to plan and perform experiments together with the analysis and interpretation of data is defined as b. Similarly, e represents the skill of students to categorize, articulate, and resolve engineering problems, and k represents their ability to apply methods and ways in solving engineering problems. However, ABET integrated some of the KPIs and reduced their number to seven. In this chapter, an algorithm for quantitative assessment of student learning level is developed. Hence, if M is the set of students undertaking a course, so m_i is the frequency of students acquiring marks in a group. The total marks obtained can be presented as follows:

$$M = \{m_1, m_2, \dots, m_n\} = \sum_{i=1}^{n} m_i$$

Similarly, let AT be the set of assessment tools, such as "major and minor exams, final exam, term project, and in-class studies":

$$AT = \{a_1, a_2, \ldots, a_k\},$$

If G is the mark a student acquired in a course at the end of a semester:

$$G = \{g_1, g_2, \ldots, g_m\}, \text{and}$$

CLO_i are the course learning outcomes:

$$CLO_i = \{CLO_1, CLO_2, \ldots, CLO_n\}, \text{and}$$

SO_j are the set of student outcomes, from *a* to *k*: which is now redefined by numbers from 1 to 7.

$$SO_j = \{1, 2, 3, \ldots, 7\}.$$

Student outcomes are the main KPIs referring skills and competencies that students must have during and after graduation. As can be seen in Table 9.3, the relation of Industrial engineering courses and KPIs are shown.

In addition to this, let R be the rate of the relation between CLOs and student outcomes, depicting the maximum level of learning. $R = \max(r_{ij})$ where $r_{ij} = 1$, 2, or 3 showing the strength of relations between SO_j and CLOs; therefore, it is an integer value and shows the level and the relevance of course learning outcomes i with program educational objectives j (student outcomes). The relevance of addressed program educational objectives is estimated in percentage by the instructor or a course design team. In this chapter, Industrial engineering program courses presented in Table 9.3, are considered for the assessment. The relevance of CLO_i(s)

Table 9.3 Contribution of course learning outcomes to student outcomes

	Contribution of course learning outcomes to student outcomes			
Course learning outcomes	*a*	*b*	*e*	*k*
CLO_1: Define quality, quality control, SQC, and TQM	3	3	3	2
CLO_2: Define TQM tools & techniques	3	3	3	3
CLO_3: Discuss the fundamentals of statistics & probability	3	3	3	3
CLO_4: Apply and analyze the variable control charts	3	3	3	3
CLO_5: Apply and analyze the control charts for attributes	3	3	3	3
CLO_6: Develop and design the acceptance sampling by attributes	2	2	3	3
CLO_7: Determine the sampling plan by different methods	2	2	3	3

(course learning outcomes) and student outcomes $SO_j(s) = \{a,...,k\}$ are presented in percentages (s_i). The design team determined the weight of relevance in percentage sequentially, as follows.

$$s_i = \{0.2, 0.3, 0.3, 0.2\}$$

$$SO_{ij} = \sum_{i=1}^{k} s_i = 1$$

The contribution of CLOs to the student outcomes is obtained by multiplying maximum learning level for each outcome with relevant addressed student outcome:

$CLO_{ij} = \sum_{i=1}^{k} r_{ij} s_i, \quad k = 1,\ldots,n,$ where n is the number of course learning outcomes.

The average class performance ($\overline{X}_i$) was calculated by the total final grade obtained, where m_i is the frequency of students who obtained grade and g_i is the mark obtained by a student:

$$\overline{X}_i = \frac{\sum_{i=1}^{n} m_i g_i}{\sum_{i=1}^{n} m_i}$$

The student learning level achieved (LOL) is calculated in percentage by multiplying course learning outcome and average class performance:

$$\text{Student level of learning}\left(LOL_i\right) = \left(CLO_{ij}\right)/100.$$

Hence, student learning level is calculated as a learning index. The learning index presents the contribution of a course to the student outcomes. Moreover, it also depicts the overall learning-level achievement of students. In engineering programs, the program-targeted level of learning is usually between [60, 100], which shows the minimum and maximum learning levels.

Evaluation of student learning starts with the development of well-established course learning objectives and student outcomes relations. Weighting of learning objectives can be modified according to the overall performance of students; therefore, questions in the exams carry much importance. On the other hand, the course design team's assessment is important to decide for the weights of learning objectives. In Table 9.3, the articulation matrices representing the distribution of CLOs into corresponding student outcomes are depicted. However, these distributions are subjective and may change by different evaluators if the course designer establishes new relations level between CLOs and SOs. In this study, the levels (strength) of learning are denoted by one, two, and three which shows average, high, and outstanding level of relation, respectively. For instance, if the relation is established based one Industrial quality control course; CLO_5 stands for the application of subject called as variable control charts. The contribution of this CLO to SO is depicted by a; ability of a student to apply math, science and engineering needs to be advanced level, hence it is depicted by 3, which means that this student must learn how to apply the control charts and show his/her ability applying the knowledge of math and engineering at advanced level. The depth of understanding at this level is the skill to synthesis or evaluate the knowledge and show competency of students. Similarly, KPI *k* shows the ability of a student to use the techniques, skills, and modern engineering tools necessary for engineering practice; CLO_1 is to define quality, quality control, statistical quality control, and total quality control. The learning level of student can be "2," which means that this student can learn at intermediate level, and the depth of understanding can be at the application and analysis level. The relationship and contribution of all CLOs with other student outcomes can be explained in a similar way. The weights of SO_i have to be determined.

As can be seen in Table 9.3, the contributions of CLOs to the student outcome a occur at different level of learning. The contributions of all CLOs for learning and applying knowledge of mathematics, science, and engineering appeared to be high for students who were enrolled in industrial quality control. In Table 9.4, the weightage of addressed student educational objectives estimated by an instructor is shown, and it is the weights designed by the instructors or course design team.

As the maximum level of learning for each student outcome is 3, the contribution of course learning outcome is obtained by multiplying the maximum level of learning for each student outcome with the relevance of addressed student outcome. In Table 9.5, the total contribution of CLOs to each program educational objective is presented.

Table 9.4 Weight of addressed student outcome in percentage

	a	*b*	*e*	*k*	Total
Relevance of addressed POs for the course percentage	20	30	30	20	100

Table 9.5 Total contribution of CLOs to each program outcome

	a	*b*	*e*	*k*
Contribution of CLOs to each program outcome	60	90	90	60

On the contrary, the average class performance $\left(\overline{X}\right)$ is calculated on the basis of the total final grade. As it was stated, m_i is the number of students obtaining grades and g_i is the numerical value of grades. Therefore, as seen in Table 9.6, the course final grades and the achievements of students in Industrial Quality Control course are presented.

If X_i is the grade obtained by a student, it can be calculated as follows:

$$\overline{X} = \frac{2472}{36} = 68.67$$

As seen in Table 9.7, the achievement of students calculated by multiplying the contribution of CLOs to SOs and average class performance are presented.

As it was stated, the Industrial Quality Control course was considered for the assessment of learning level. The number of students enrolled in this course was 36 for spring 2012 semester. The number of students obtaining grades A+ to D was 30. Therefore, as seen in Table 9.8, the first row shows the program achievement level of this course. The course supports four SOs and there are five more courses supporting the same program educational objectives whose achievements are presented in Table 9.8.

As seen in Table 9.8, the courses addressing student outcomes on the basis of average course contribution are presented. These averages show the learning level for each of the student outcomes. In Table 9.9, the average

Table 9.6 Course final grades and frequency distribution

Grades	Numerical values (g_i)	Number of students obtaining grades (f_i)	Achievement
A+	95	4	380
A	90	4	360
B+	85	2	170
B	80	0	0
C+	75	7	525
C	70	1	70
D+	65	5	325
D	60	7	420
F	37	6	222
Total		36	2,472

Table 9.7 Achievement of students addressed by the contribution of CLOs to SOs

Achievement of students in percentage	41.17%	61.75%	61.75%	41.17%

Table 9.8 Program achievement level for six courses addressing the program outcome

Courses	*a*	*b*	*e*	*k*	Program-targeted max. learning level	Program-targeted min. learning level
1	43.4	65	65	43.4	100 (5 in GPA)	60 (2 in GPA)
2	78.6	83.2	67.5	75.3	100 (5 in GPA)	60 (2 in GPA)
3	82.1	74.3	75.4	81.2	100 (5 in GPA)	60 (2 in GPA)
4	76.5	70.6	58.5	86.3	100 (5 in GPA)	60 (2 in GPA)
5	70.3	68.6	67.4	55.4	100 (5 in GPA)	60 (2 in GPA)
6	68.9	72.2	70.6	56.3	100 (5 in GPA)	60 (2 in GPA)

Table 9.9 Contribution of courses to program outcomes

Course contributions to program outcomes ("*a*" to "*k*"); (Achievement level)							
Courses	1	2	3	4	5	6	7
1	43.4	65.0			65.0		43.4
2	78.6	83.2			67.5		75.3
3	82.1	74.3			75.4		81.2
4	76.5	70.6			58.5		86.3
5	70.3	68.6			67.4		55.4
6	68.9	72.2			70.6		56.3
22	58.3	67.5			70.5		64.5
Average	70	72	70	72	67	52	66

contribution of 22 courses to student outcome is presented. It is realized that average contributions provide useful information about course learning level, which is low in some cases, but they can be improved by well-designed course contents, and tools supporting each student outcomes. With regard to engineering program assessment, senior-level courses provide better understanding of subjects with better overall learning performance.

Figures 9.4b and 9.5 show the overall learning assessment for the student outcome. Learning level is a quality characteristic that cannot be normally represented numerically. However, due to the current applications, the quality of learning has to be presented by numbers for understanding its assessment easily. In this chapter, we define specification limits for learning quality and have developed control charts to present the contribution of courses to the program educational objectives. Different assessment tools can be used to evaluate the learning level of students. GPA of students is one of the important indicators for their learning level, and the category that possibly they might be included during assessment.

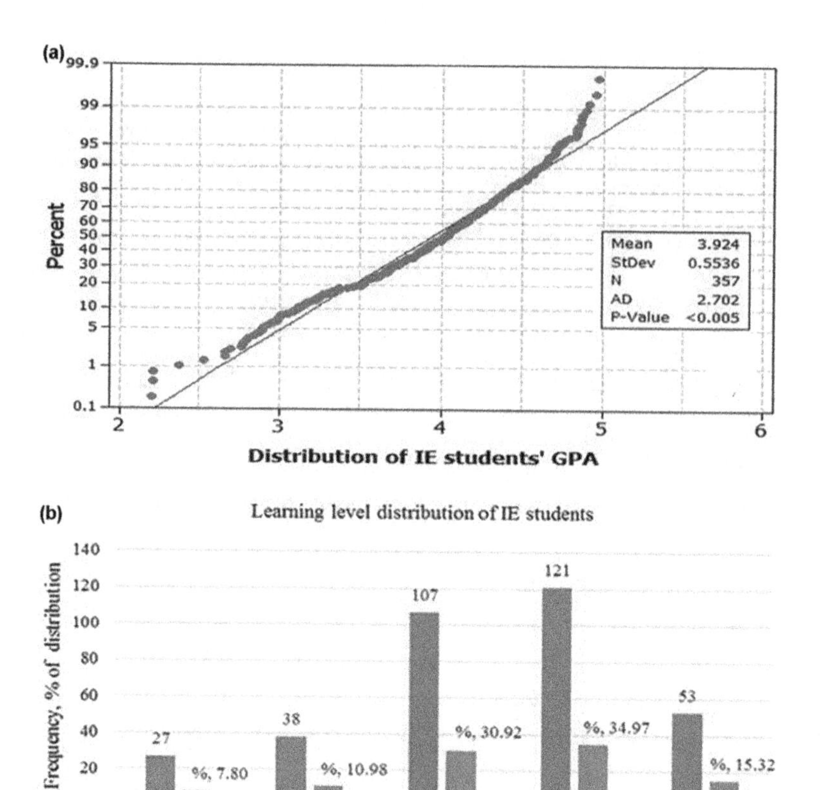

Figure 9.4 Distribution of IE student GPA (a) and the learning-level groups (b).

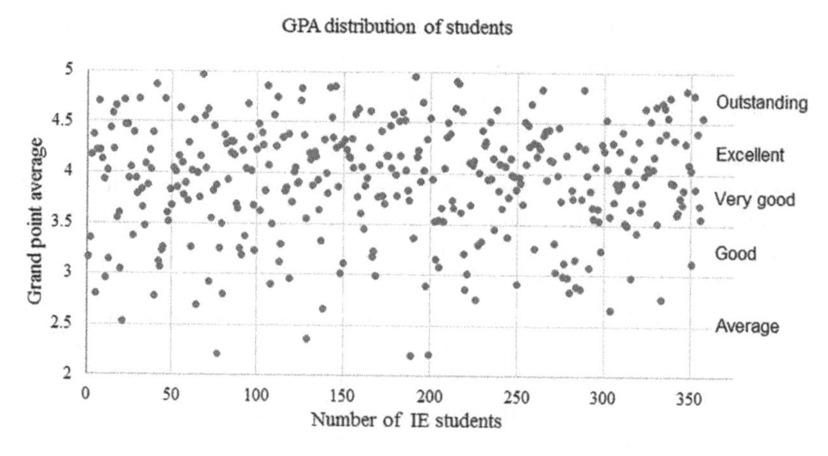

Figure 9.5 GPA of IE students and its distribution and assessment.

Law (1995) used linguistic terms for assessment of students' achievement, such as fail, average, excellent, and outstanding.

Figure 9.4a shows the distribution of GPA of IE students. This distribution is normal, the mean of distribution is 3.924, and the standard deviation is 0.5536. The *P*-value of distribution is less than 0.005, which indicates that the distribution is meaningful. Figure 9.4b depicts the groups of learning levels. These groups were established based on the GPA of 345 students graduated from industrial engineering department during last 4 years at King Abdulaziz University of Jeddah in Kingdom of Saudi Arabia. Mainly, five groups were established, and as seen in Figure 9.4b, the learning level of these groups of students has GPA between (2 and 2.99). These students are at the novice level of learning, their learning level is "average," and they are at the knowledge and comprehension level, which indicates that they are still at the introduction level of learning. These students can remember of previously learned materials which may involve the recall of a wide range of material from specific facts to complete models, and they have ability to grasp the meaning of materials by interpreting those estimating future trends. 7.8% of IE students (27 students) graduated are at this level of learning.

10.98% of students have GPA between (3.0 and 3.44), and there are 38 students in this group. These students are at the intermediate level of learning, and their learning level is good which stands for application and analysis. Students at this level can focus on the application of knowledge and analysis of findings. Engineering students at this level of learning have the ability to use the learned material in new, concrete situations. They might include the application of rules, methods, principles, and theories which require a high level of understanding. Similarly, engineering students achieving the application level of learning have ability to disintegrate material into its components so that its organizational structure can be understood. They can include the identification of information pieces, analysis of the relationship between these pieces of information parts, and recognize the organizational principles involved.

Out of 354 students, 107 students have GPA between (3.5 and 3.99) in this group which comprises about 30.92% of students. These students are also at the intermediate level of learning, but they have "very good" level of learning which stands for analysis of knowledge. Students at this level can focus on the analysis of knowledge and their findings. Industrial engineering students at this level of learning have the ability to use the learned engineering tools in developing new and concrete situations. They apply methods, principles, and theories which require a higher level of understanding. Similarly, these groups of students have achieved the ability to disintegrate information into components, and can include the information pieces in the whole, analyze the relationship between the pieces of information parts, and recognize the organizational principles involved.

Similarly, 34.97% of students have GPA between 4.0 and 4.44, and 121 students are in this group of learning level. These students are at the advanced level of learning, and their learning level is excellent which stands for synthesis and evaluation level of learning. For instance, at the "synthesis" level of learning, engineering students graduated have ability to put parts of information together to form a new whole, and have good communication skill, can make a plan for operations, or establish a set of abstract relations for classifying information and develop some more details. These students can also illustrate creative behaviors with major emphasis on the formulation of new patterns or structure. Together with synthesis, students at this level of learning can evaluate information, and have ability to judge the value of material for a given purpose based on definite criteria which contain elements of all the other categories, and judgments based on clearly defined assessment criteria.

On the other hand, 15.31% of students have GPA between (4.5 and 5.0), and 53 students are in this group of learning level. These students have advanced level of learning, and they have achieved deep learning level. Their learning level is outstanding which stands for synthesis and evaluation ability of learning. Students at this level can focus on the application of knowledge, merging pieces of information to establish a whole and analysis of findings. They have achieved deep learning level, their technical and soft skills are quite high, and their competency and capability of learning are also higher. They can apply rules, methods, and theories quite easily. Their synthesis capability is better than that of the other groups, and their competency-based learning level is outstanding. They can disintegrate material into its components, and include the identification of information pieces, and analysis the relationship between these pieces. Figure 9.5 depicts the GPA of IE students, the distribution of learning level, and its assessment. Figure 9.5 also depicts the total effects of all industrial engineering courses on the learning level of students, and if the content of courses is carefully selected, the technical and soft skills of students will be higher. Students' competency-based learning level will also be effective in their learning capability.

One of the basic objectives of these assessment methods is to ensure effective and efficient learning (Taylan et al., 2017). As an evident in a simple curving system, the instructor determines beforehand that very few students will do either very well to get "outstanding" or very poor to fail, whereas the majority will cluster in the middle to be "good," "very good" or "excellent." Figure 9.4a shows normal probability plot and grade distribution of learning performance for a real data set obtained from the outcomes of students who have taken industrial quality control course. On the other hand, Figures 9.6a and b show the desirability of students' learning level which illustrates the performance of two groups of students. The mean of each group is the same; however, the standard deviations are different. It is clear that when the standard deviation is large, the shape of the curve is platy-kurtic and when the standard deviation of the learning level is small, the curve is leptokurtic.

Some instructors might not consider the distinction between the two cases in terms of students' learning quality level; that is, in both cases, virtually all students are in the passing limits. However, the students represent two different learning levels; a clear distinction exists between the students in Figure 9.6a and b with regard to quality of learning. For

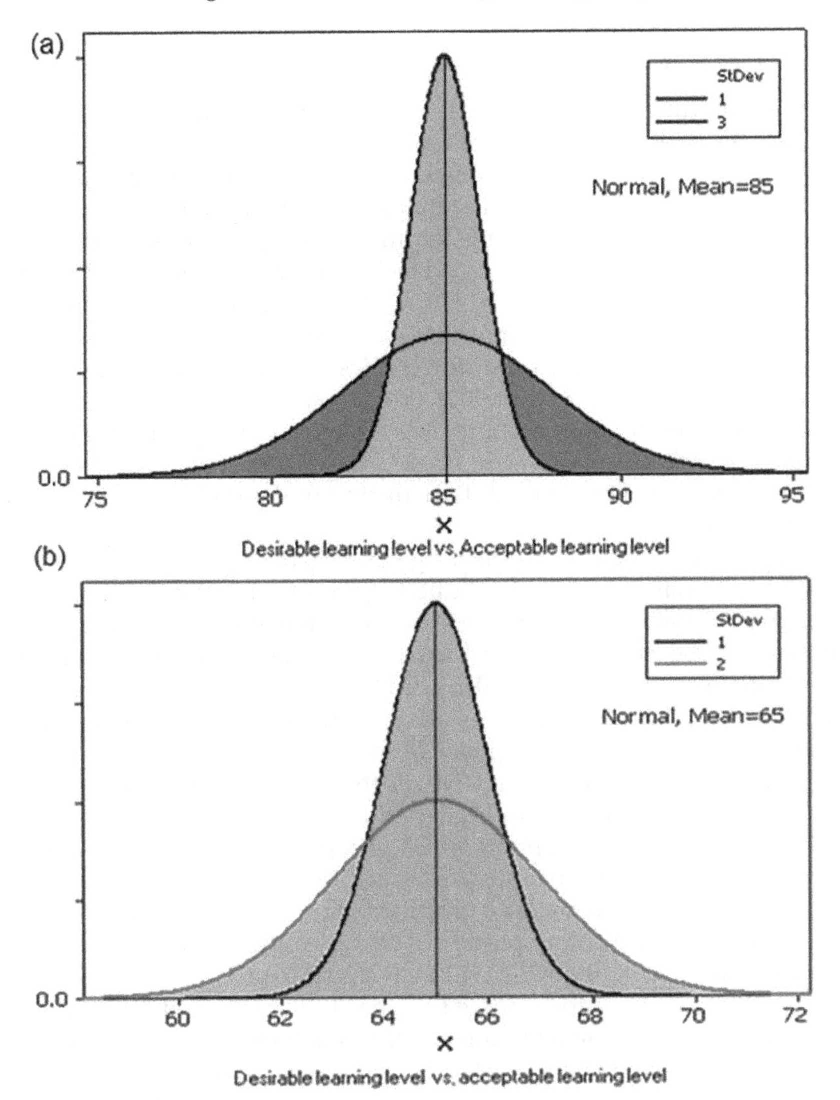

Figure 9.6 Desirability and acceptability of learning level (Taylan et al., 2017).

instance, an instructor analyzing the exam performance of students would categorize two students with scores of 58 and 62 as the same, but much poorer from those who got 85 or 95. Even though 60 is the lower limit for passing grade, no one would view the two students with grades of 62 and 95 as the same from a quality standpoint (Taylan et al., 2017). This low-level learning may not only have harmful effects in the future work life of students' but also bring higher cost to the society in terms of poor work quality and low productivity.

9.3 Conclusion

Student learning level is strongly dependent on course's design and assessment tools. Before or after graduation, students must have technical and soft skills enough to meet the diverse needs of the growing and changing demands of global industries. The technical skills might be obtained from the knowledge acquired during undertaking courses. There is a rising need for the skilled technical staff across several industries in the globe. Competency-based student learning might help to meet those needs of industry. However, some other problems might become a current issue in the universities. It is necessary to ask here several questions: Can universities provide all technical and soft skills that students need in the real life? The answer of this question is not so simple. If cannot, then who will provide these skills? If yes, who is going to pay for it or is there enough time during undergraduate program to cover everything?

Therefore, design of engineering courses, their contents, and some other additional programs might be helpful to achieve certain level of learning. Well-designed learning objectives together with course contents should be clearly explained to students, and reciprocally, students' expectations from instructors also need to be properly addressed. In this chapter, three learning domains are systematically scrutinized. These domains are designed as the house of cognitive learning to explore students' different learning levels which they are supposed to attain in their engineering courses. In higher education system, the assessment of deep learning level is an important phenomenon. Therefore, an integrated integer-programming approach is developed to assess the correct learning level of students. The proposed algorithm assessed the learning quality of students and presented the outcomes as a learning index.

As a conclusion, the KPIs will help students take more responsibility of their studies and determine their needs and the needs of industry. Students are expected to do frequent self-evaluations on their competency and feel more confident in developing their own plans for a better learning. Continuous learning and improvement tools can be adopted to curriculum design to achieve the planned goals. On the contrary, educational

IT infrastructure can be developed to encourage students to submit their scholarly works on- or off-campus and prepare them for interactive discussions during classes.

References

ABET (2016). Retrieved from www.abet.org/accreditation/accreditation-criteria/accreditation- policy-and-procedure-manual-appm-2016-2017.

Anderson, L. W., & Krathwohl, D. (Eds.) (2001). A taxonomy for learning, teaching, and assessing: A revision of Bloom's taxonomy of educational objectives. Retrieved from http://thesecondprinciple.com/optimal-learning/brainbased-education-an-overview/.

Baird, J. A. (2013). Judging students' performances. *Assessment in Education: Principles, Policy & Practice*, 20, 247–249.

Ball, A. G., Zaugg, H., Davies, R., Tateishi, I., Parkinson, A. R., Jensen, C. G., & Magleby, S. P. (2012). Identification and validation of a set of global competencies for engineering students. *International Journal of Engineering Education*, 28(1), 156–168.

Black, P. (2015). Formative assessment—An optimistic but incomplete vision. *Assessment in Education: Principles, Policy & Practice*, 22, 161–177.

Bloom, B. S. (1956). *Taxonomy of Educational Objectives, Handbook I: The Cognitive Domain*. New York: David McKay Co Inc.

Bloom, B. S. (1984). The 2 sigma problem: The search for methods of group instruction as effective as one-to-one tutoring. *Educational Researcher*, 13(6), 4–16.

Brookhart, S. M. (2013). The use of teacher judgement for summative assessment in the USA. *Assessment in Education: Principles, Policy & Practice*, 20, 69–90.

Dadach, Z. E. (2013). Quantifying the effect of an active learning strategy on the motivation of students. *International Journal of Engineering Education*, 29, 904–913.

Deng, H., Yeh C.-H., & Willis, R. J. (2000). Inter-company comparison using modified TOPSIS with objective weights. *Computers & Operations Research*, 27, 963–973.

Dominguez, C., Nascimento, C., Carreira, R. P., Cruz, G., Silva, H., Lopes, J., ..., Morais, E. (2014). Adding value to the learning process by online peer review activities: Towards the elaboration of a methodology to promote critical thinking in future engineers. *European Journal of Engineering Education*. Advance online publication. doi:10.1080/03043797.2014.987649.

European Commission Using Student outcomes (2011). European Qualification Framework Series: Note 4, Publications Office of the European Union, Luxembourg, 1–48.

Freeman, R., & Dobbins, K. (2013). Are we serious about enhancing courses? Using the principles of assessment for learning to enhance course evaluation. *Assessment & Evaluation in Higher Education*, 38, 142–151.

Grez, L. D., & Valcke, M. (2013). Student response system and how to make engineering students learn oral presentation skills. *International Journal of Engineering Education*, 29, 940–947.

Grissom, J. A., & Loeb, D. K. S. (2015). Using student test scores to measure principal performance. *Educational Evaluation and Policy Analysis*, 37, 3–28.

Hargreaves, E. (2007). The validity of collaborative assessment for learning. *Assessment in Education: Principles, Policy & Practice*, 14, 185–199.

Hayward, L. (2015). Assessment is learning: The preposition vanishes. *Assessment in Education: Principles, Policy & Practice*, 22, 27–43.
Klenowski, V., Askew, S., & Carnell, E. (2006). Portfolios for learning, assessment and professional development in higher education. *Assessment & Evaluation in Higher Education*, 31, 267–286.
Krathwohl, D. R. (2002). A revision of Bloom's taxonomy: An overview. *Theory into Practice (Routledge)*, 41, 212–218.
Law, C. K. (1995). Using fuzzy numbers in educational grading system. *Fuzzy Set and Systems*, 83, 311–323.
Liu, O. L., Frankel, L., & Roohr, K. C. (2014). Assessing critical thinking in higher education: Current state and directions for next-generation assessment. *ETS Research Report Series*, 1, 1–23.
Lopes, A. L. M., Lanzer, E. A., & Barcia, R. M. (1997). Fuzzy cross-evaluation of the performance of academic departments within a university. *Proceedings of the Canadian Institutional Research and Planning Association Conference*, Toronto, Canada, October, 19–21.
Ma, J., & Zhou, D. (2000) Fuzzy set approach to the assessment of student-centered learning. *IEEE Transactions on Education*, 43(2), 237–241.
Meyer, J. H. F., Knight, D. B., Callaghan, D. P., & Baldock, T. E. (2014). An empirical exploration of metacognitive assessment activities in a third-year civil engineering hydraulics course. *European Journal of Engineering Education*. Advance online publication. doi:10.1080/03043797.2014.96036.
Northwood, D. O. (2013). Student outcomes—some reflections on their value and potential drawbacks. *World Transactions on Engineering and Technology Education*, 11, 137–142.
Palermo, J. (2011). Linking student evaluations to institutional goals: A change story. *Assessment & Evaluation in Higher Education*, 38, 211–223.
Polikoff, M. S., & Porter, A. C. (2014). Instructional alignment as a measure of teaching quality. *Educational Evaluation and Policy Analysis*, 36, 399–416.
Reich, A., Rooney, D., Gardner, A., Willey, K., Boud, D., & Fitzgerald, T. (2014). Engineers' professional learning: A practice-theory perspective. *European Journal of Engineering Education*. Advance online publication. doi:10.1080/03043797.2014.967181.
Rogers, G. (2003). Do grades make the grade for program assessment? Retrieved from www.uri.edu. Self-paces, open-entry/open-exit/ format. *Paper presented at the ASEE Annual Conference and Exposition*, Seattle, WA.
Spelt, E. J. H., Luning, P. A., Boekel, S., & Mulder, M. (2014). Constructively aligned teaching and learning in higher education in engineering: What do students perceive as contributing to the learning of interdisciplinary thinking? *European Journal of Engineering Education*. Advance online publication. doi:10.1080/03043797.2014.987647.
Tamburri, R. (2013). Trend to measure Student outcomes gains proponents, Canadian University Affair, Toronto, February 26, Article.
Taylan, O., Bafail, A. O., Abdulaal, R. M. S., & Kabli, M. R. (2014). Construction projects selection and risk assessment by fuzzy AHP and fuzzy TOPSIS methodologies. *Applied Soft Computing Journal*. 17, 105–116.
Taylan, O., Ridwan, A., & Parsaei, H. (2017). Assessment of student learning by hybrid methods at program level. Engineering Education Letters, QScience, 1–14.

Taylan, O., & Karagozoglu, B. (2009). An adaptive neuro-fuzzy model for prediction of student's academic performance. *Computers & Industrial Engineering*, 57, 732–741.

Tian, M., & Lowe, J. (2013). The role of feedback in cross-cultural learning: A case study of Chinese taught postgraduate students in a UK university. *Assessment & Evaluation in Higher Education*, 38, 580–598.

Wang, C. C., Kang, Y., Chang, Y. J., & Chang, Y. P. (2007). Application of fuzzy neural networks for grading. *25th IASTED International Multi Conference, Artificial Intelligence & Applications*, 12, 78–83.

chapter ten

From "The Academy" and "Lions' Circle" to "NiTiM"

A success story of advanced study programs

Maximilian Moll, Marian Sorin Nistor, and Stefan W. Pickl
Universität der Bundeswehr München, COMTESSA

Contents

10.1 "The Academy": Beginning of a new journey

> Every student can learn - just not in the same way or on the same day
>
> **George Evans**

This fact is widely recognized, but usually only seen from one side. As a consequence, a lot of support and understanding was built for children struggling at school. The fact that the same principle can be applied to the upper part of the spectrum was ignored for a long time. Nevertheless, a regular school can be almost as frustrating to a gifted as to a struggling student. Fortunately, this situation started to improve during the last decade with much more awareness being raised. In Bavaria, Germany, for example, it even got to the point of assigning a teacher at every school to be responsible for and of help to gifted students. Furthermore, there are now programs within schools, hosted by governmental and nongovernmental institutions and universities to provide additional opportunities to extend advanced learning beyond the school classroom.

Most gifted programs at universities are built around the idea of students participating in lectures during the last few years of their school education.

The exact program varies between universities, but usually they either have specially designed and taught lectures or join the regular lectures. This system has some benefits, most prominently to shorten later studies by being able to accredit courses taken during this time. We called the corresponding program at the Bundeswehr University Munich "The Academy" to signify that we aimed for more. Being the first program of this kind in Southern Germany, students from the age of 14 with an outstanding overall performance across all school subjects were selected. This requirement is in contrast to many similar programs, for which it is sufficient to excel just in the subject they want to study. Additionally, they were required to provide a letter of recommendation from their school and a parental agreement.

Furthermore, we expected them also to show significant extra-curricular activities, which we used as an indicator, that they did not have to work too hard to achieve excellence at school. It is interesting to note that we did not ask for a classical IQ-test since the emphasis was placed on educating personalities. Furthermore, the other requirements established their aptitude sufficiently, making further tests redundant. In exchange, we offered them to take any course they wanted, guided by personal counseling, instead of limiting them to a single field of studies.

Moreover, we had a trained psychologist and support from the coordinator and the supervising professor available to make sure that we could help them handle any problems arising on any level. Overall, this tries to achieve the declared goal of the program to coach students in their entirety by supporting and systematically challenge them.

10.2 The "Lions' Circle": A promising sequel

With the "Lions' Circle" - more on the name will follow later - we took a very different approach. Nevertheless, the underlying idea remains the same:

> Creating additional opportunities for talented students by exposing them to university level mathematics.

However, we wanted to move away from the fixed setting of frontal teaching, and develop a unique program that can provide a dynamic environment that facilitates a growth toward being independent problem solvers instead. In other words, the primary goal was not just to give them new information, but to teach them how to acquire new knowledge and to approach problems on their own. The ability to approach problems well is essential for mathematicians, computer scientists, and engineers and provides a particular problem for good students since they are often not challenged at school. This typically leads to a deprecation of their ability to ask good questions. Hence, much emphasis was placed on developing this skill. In a first step, any heads-on teaching was presented at comparatively high speed. This conditioned them first and foremost to ask questions to slow down the lecturer. At that point though, most questions where directed back at the group. Trying to find an answer made not only the rest of the group understand, why the question was asked, but also gave them feedback on the quality of the question. Over time, significant improvement could be observed. Often though, new topics or pieces of knowledge were introduced by highlighting an obvious problem and let the members of the group try to discover the solution themselves. This approach, while not feasible in almost all other settings of teaching, raises much more understanding for the background of the knowledge acquired. This discovery process was held often as group discussion. Even more beneficial, though, seems to be splitting them up into smaller groups for brainstorming first (see Figure 10.1). Often it could be observed that each group reached certain hurdles at pretty much the same time. However, different approaches between them led to fruitful discussions afterward, when comparing their work.

10.3 Reactions to a new environment

This work in small groups, but also the whole program in general, gives them also a chance to meet and interact with like-minded people from their age group. For many of them, this is the first time they are not the best mathematician in the room. This requires them also to grow personally to learn to deal with that. It is interesting to observe that there are mainly three different types of personalities present.

The first type realizes his or her change in status and is acutely afraid of looking out of place in front of the other "smart" participants. This often

Figure 10.1 Small group for brainstorming at "Lions' Circle." (Copyright of Hochschulkurier, 2017.)

leads to them being quiet and reluctant to participate. The challenge here is to make them lose this fear by generating an open and friendly atmosphere.

The second type starts to flourish in this surrounding, and enjoys the discussions and the intellectual challenge that comes with it.

Finally, there is the occasional student, who prefers to quietly sit in the classroom and more or less does his or her own stuff, while simultaneously paying attention. The choice made here was to mostly let them do their things since they still seemed to benefit.

To further increase problem-solving capabilities, riddles were posed often to be solved until the next session. One of these led to the name for the program:

> A lion can only move on the circumference of a circle, at the center of which is a man.
>
> The lion can run four times as fast as the man.
>
> How can the man, who can move freely inside the circle, escape,
>
> even if the lion chooses optimal behavior at every point in time?

10.4 A Holistic Approach to teaching advanced Operations Research at "Lions' Circle"

Thematically, mathematics was considered from three different viewpoints: theory, algorithms, and implementation. The theory part focused on traditional pure maths, as it would be lectured at university. It made sure to introduce standard notation and put much emphasis on learning to find proofs for theorems. Within the algorithms part, this theory was then used to propose and study algorithms with a focus on understanding exactly how they work and a very hands-on approach. Nevertheless, the connection to the theoretical footing was always kept in mind. Finally, for the implementation part, the focus was on Python and using it to implement the algorithms studied. This made the program also approachable for students more interested in computer science than mathematics.

The first major topic was linear programming. For the theory part, it could be used to introduce basic notation, some fundamental linear algebra, and fairly simple proofs with a clear structure. The main algorithm was the Simplex algorithm, which was preceded by Gaussian Elimination. This choice was not only sensible didactically but also gave them a tool, which they could use for their final exams at school. Before turning to the implementation of both algorithms, Python needed to be introduced, but the focus on programming as a mean for implementing mathematical algorithms was kept throughout. Further topics were then discussed with the group, and the next choice landed on machine learning. Here, neural networks and reinforcement learning was discussed, so that the group was capable of designing a program, which could learn to play a game of their choice. A logical choice after this was to turn to data science in general, since its importance is ever growing, and it allows the participants to demonstrate a fair amount of creativity.

The main issue with this form of program cannot be avoided. The students were selected by good marks at school and how much extra-curricular interests and activities they demonstrated. A direct consequence is that they have quite full timetables, making it difficult to find a time slot suitable for everyone every time. This leads to a certain amount of fluctuation of students present, which hinders continuity slightly. However, this was never a serious issue.

10.5 The "NiTiM" International Graduate School for Research

Graduate schools for research are generally nonprofit entities that bring researchers from academia and industry together to exchange ideas, knowledge, and experience on specific topics. While they are affiliated with various departments like colleges, universities, and industrial partners, they are considered independent organizations.

The classical graduate schools are just an advanced form of education with courses and programs tailored to their respective field of studies. They usually come with a rigorous curriculum, for which universities and colleges often impose their own institutional standards (Katz, 1966). Graduate schools for research, on the other hand, are focusing on supporting the careers of young researchers, tailored more to the individual achievements and goals, in other words being an extension of the formal studies and not an alternative to them.

Admission to a graduate school for research is also not strictly requiring students to have graduated but is decided based on their level of research experience. As a relative reference point, the education of a student during bachelor's studies focuses approximatively 80% on building a foundation of the studied field, and only 20% on research. For master's studies, the first level of graduate students, the percentage is vice versa. Since the speed of assimilation for students varies, advanced bachelor's students can already enroll in graduate schools of research. One of the main advantages of these schools is bringing a certain level of quality of research across multiple institutions. Ph.D. programs are highly specific to every institution—even universities in the same city can have different structures and standards for a Ph.D. program. Bringing together members of various institutions under one umbrella can establish a certain standard for work ethic and research.

This is also the purpose of the International Graduate School for Research on Networks, Information Technology and Innovation Management, or shorter "NiTiM" (Universität der Bundeswehr München, 2018). A formal nonprofit organization brings together various disciplines like product management, business informatics, operations research, entrepreneurship, and others to collaborate on specific topics like innovation management, crisis management, and so on.

It was formed by a group of professors that met through their Ph.D. program at the University of St. Gallen, Switzerland. Through this school, they decided to pass their educational standards to their Ph.D. candidates, independent of the institution, with which they continued their career. Thus, professors now in Germany, The Netherlands, Spain, and Italy, brought together graduate students in an international, intercultural, interdisciplinary environment for an academic and industrial exchange.

A graduate school like NiTiM does not formally graduate students. Its members are already enrolled at accredited institutions for this process. The role of such a school is to offer a research set-up and quite often to apply for research funds (European Commission, 2017) to encourage young researchers to develop a career in their field of studies. Since 2005 NiTiM is helping students to graduate at their institutions by offering additional ways to grow: doctoral seminars, collaborative teams, doctoral schools, secondments, teaching/mentoring/supervising opportunities, career development plans, and others.

10.6 International Doctoral Seminars at NiTiM

A NiTiM doctoral seminar, also called regional learning circle, brings together graduate students from various institutions which are somewhat geographically close located. NiTiM holds three parallel seminars in Munich-Germany, Barcelona-Spain, and Leiden (The Netherlands/Münster-Germany).

Here, graduate students meet once a month to share their latest research progress and to get feedback on their work. Once in a while, everyone gets the chance to defend his or her work in front of peers and supervisors in an attempt to support the presenter's direction of work.

As a small parenthesis, it is known that a low confidence level on students' personal research development with the lack of support from supervisors and peers is considered, after competence reasons, a critical factor of studies dropout (Litalien and Guay, 2015).

Another role of these seminars is to offer a good environment to practice talks for scientific presentation, benefit from direct feedback, and also to improve the language skills of non-native speakers. Homework is also given in the form of paper reviews. The graduate students can offer their ongoing research papers to peers for feedback, or they are assigned various papers from the desired field of research to improve their reviewing skills. Every few months, experts or practitioners are invited to share their work experience with the group. Small workshops are hosted twice a year on topics like "Presentation with Impact" (Berman and Cheng, 2010) and "Academic Writing" (Swales and Feak, 2004).

The collaborative teams are self-organized groups of graduate students that address specific topics with the support of faculty members.

At NiTiM there are three such teams: the School Organizing Team, the Editorial Board, and the Collaboration Team.

10.7 The NiTiM School Organizing Team

The organizing team consists of four graduate students plus one local faculty member. The Chair is responsible for the primary organization, supported by the Outgoing Chair, who organized the previous school and provides experience exchange. The Incoming Chair, who will organize the next school, is joining as well, to have enough time to get to know the organization process.

They are further supported by the right-hand of the Chair, the Program Chair, and the Local Chair, which is a local faculty member that will help with the administrative tasks at the organizing institution.

This team organizes every year two doctoral schools and has a rotational system of members to offer each graduate student at least one organizing experience.

The role of doctoral schools is to gather as many NiTiM members as possible for a 3-day experience at one of the partner institutions or another venue shared with a conference from the same field of research. This school, which can be considered a more focused version of the NiTiM seminars, is bringing all groups together with possibly additional guests. Since the number of participants is limited, a selection of graduate students is made in a process similar to conferences. The graduate students submit their current research proposal to the school. Each submission is reviewed by at least two professors from the field, and the best ones are selected for presentation during the school. The primary purpose of the doctoral school is to give a chance to each selected candidate to present and receive feedback for one and a half hours from a commission of professors and peers from the same (or similarly appropriate) fields of research. Also, plenary talks, thematic workshops, and social events are organized, very similar in structure to a regular conference.

10.8 The NiTiM Editorial Board and Collaboration Teams

The Editorial Board is formed by four graduate students who are publishing monthly NiTiM newsletters. Their roles are split into three editors and a coordinator, the editor-in-chief. The purpose of this newsletter is to inform the entire graduate school (and its contacts' network) about various activities like incoming events, collaboration opportunities, recent publications from the network members, and it serves as a platform for graduate students to share their experience after attending various training programs or conferences. The roles are not fixed, like in the case of the other team, which offers each member a 6-month experience as an editor. The participation in the Editorial Board team brings various benefits, from improving written communication skills up to networking opportunities. This team is closely connected with the Collaboration Team. The latter is a team of three graduate students who are administrating the collaboration platform of the graduate school. Each member represents one regional learning circle, and their role is to upload internal papers, the research proposals that were presented during the doctoral schools, the minutes of NiTiM seminars, the career development plans, and other research-related documents to the NiTiM collaboration platform. Here, the roles are assigned voluntarily and usually are accepted by candidates with technical skills. A platform like this is vital for a graduate school to gather all the research together in one place.

10.9 The NiTiM career development plan and further opportunities

At NiTiM every member is holding a career development plan. In the beginning, the candidates are asked to write the main direction of their career they pursue, their current achievements and skills, and the ones they would aim to achieve at NiTiM with the support of their supervisors. These plans are revised once a year according to the current progress and adjusted in case the aim has changed in the meantime. This is a very effective tool to not get sidetracked and to better focus on one's goals.

The opportunity for secondments is something that not every graduate school has. NiTiM offers through its vast network of affiliated partners the possibility of academic or industry exchange called secondment. For example, not every Ph.D. topic is straightforward and only library connected.

In case there is the need for a specific expert knowledge that is hard to reach at the current institution or data collection in the field, the supervisor can send their Ph.D. candidate for a determined period to another university or industrial partner within the worldwide NiTiM affiliation network.

Furthermore, there is also the possibility of mixed events. NiTiM members are encouraged to collaborate with other groups for activities on more specific research directions. One example is the competence center "COMTESSA" affiliated with the House of Logistics and Mobility (HOLM) at Frankfurt Airport, Germany. Here, COMTESSA organizes every year a Ph.D. seminar in a very creative atmosphere. One of the opportunities for participants is to exchange ideas with practitioners about applied research directions. This is one way how COMTESSA encourages ideas from industry to inspire exciting research topics in the field of Operations Research.

Finally, most academic institutions take for granted that their Ph.D. candidates have the opportunity to gather experience in soft skills like teaching, mentoring, or supervising undergraduate students for seminars, laboratories, examples classes, and others.

This is a significant disadvantage for the Ph.D. candidates from the industry that would like to consider an academic career in the future. A similar problem arises for international candidates at colleges or universities that don't offer to teach in English. NiTiM offers agreements between institutions to overcome this issue.

10.10 Conclusions

Working with pupils and students is bringing a particular satisfaction of appreciation for having the chance to contribute to their educational training. Any of the mentioned programs in this chapter have advantages and disadvantages since there is no one-method-fits-all approach. Therefore, it is an educational training for the teachers and professors as well, as each experience contributes to a better approach for the next generation.

The response to the first two programs was overall very good. Even students having to leave these programs did so for different reasons, regretting having to do so. It would be good to see more programs of this kind since it is gratifying and challenging to the lecturer. However, it also raises impressive future researchers, who might have otherwise dulled down at school.

The NiTiM graduate school for research also has good results, increasing the ratio of successful Ph.D. programs by keeping the candidates motivated, periodically trained, and encouraged to pursue activities and to attend events that can contribute to their career development plan. The bi-annual Ph.D. proposal presentation was found to be the most suitable method to measure the research progress and to adapt the next steps accordingly to, e.g., avoid research procrastination as much as possible.

This chapter intends to stimulate open discussions not only between professors and students but also among professors and teachers to better follow the needs of every new generation of students. This can increase the number of tailored programs that could keep the students motivated to finish the studies and to start with confidence a career based on what they learned during their educational training.

In the memory of Prof. Dr. Bernhard Katzy.

References

Berman, Robert; Cheng, Liying (2010). English academic language skills: Perceived difficulties by undergraduate and graduate students, and their academic achievement. *Canadian Journal of Applied Linguistics* 4 (1), 25–40.

European Commission (2017). A Networked and IT-Enabled Firm's Perspective on Crisis Management. Available online at https://cordis.europa.eu/project/rcn/105344_en.html, updated on 9/24/2017, checked on 11/13/2018.

Hochschulkurier (2017). Einblicke in die Welt der Wissenschaft: Willkommen im Mathe-Club. Nr.59 | July 2017.

Katz, Michael B. (1966). From theory to survey in graduate schools of education. *The Journal of Higher Education* 37 (6), 325–334.

Litalien, David; Guay, Frédéric (2015). Dropout intentions in PhD studies: A comprehensive model based on interpersonal relationships and motivational resources. *Contemporary Educational Psychology* 41, 218–231.

Swales, John M.; Feak, Christine B. (2004). *Academic Writing for Graduate Students: Essential Tasks and Skills*; University of Michigan Press: Ann Arbor, MI.

Universität der Bundeswehr München (2018). Cooperation on Research and Innovation. Available online at https://www.unibw.de/forschung-en, checked on 11/13/2018.

Index